Troubleshooting and
Root Cause Failure Analysis

Troubleshooting and Root Cause Failure Analysis

Equipment Problem Solving

J. R. Paul Lanthier, P. Eng.

INDUSTRIAL PRESS

Industrial Press, Inc.

32 Haviland Street, Suite 3
South Norwalk, Connecticut 06854
Phone: 203-956-5593
Toll-Free in USA: 888-528-7852
Fax: 203-354-9391
Email: info@industrialpress.com

Author: J. R. Paul Lanthier, P. Eng.
Title: Troubleshooting and Root Cause Failure Analysis: Equipment
Problem Solving
Library of Congress Control Number: 2020950418

© by Industrial Press, Inc.
All rights reserved. Published in 2021.
Printed in the United States of America.

ISBN (print):   978-0-8311-3665-9
ISBN (ePUB):   978-0-8311-9589-2
ISBN (eMOBI):   978-0-8311-9590-8
ISBN (ePDF):   978-0-8311-9588-5

Publisher/Editorial Director: Judy Bass
Copy Editor: Elise Davies
Compositor: Patricia Wallenburg, TypeWriting
Proofreader: Claire Splan
Indexer: Claire Splan

industrialpress.com
books.industrialpress.com
1 2 3 4 5 6 7 8 9 10

Contents

CHAPTER 1

Introduction to Problem Solving   1

PART I

UNDERSTAND FAILURES

CHAPTER 2

Understanding Failures   21

PART II

TROUBLESHOOTING

PART III

ROOT CAUSE FAILURE ANALYSIS (RCFA)

CHAPTER 6

Conclusion   185

Glossary   189

List of Figures

List of Tables

Troubleshooting and
Root Cause Failure Analysis

Introduction to Problem Solving

Problem solving is the act of recognizing a problem, determining its cause(s), and identifying and implementing the appropriate solution(s) to resolve (or at least mitigate its consequences). In industry, good problem-solving techniques and skills help us solve issues efficiently and effectively and maintain or resume control of our assets and work environment while minimizing the impact on safety, the environment, quality, throughput, customer service, and cost.

Progressive organizations recognize the impact of effective problem solving on the performance of their site and establish programs to develop these skills and techniques within their workforce. There are two basic types of problem-solving techniques:

1. **Troubleshooting.** Focused on resolving the effects quickly so that we can go back to normal, or *near-normal*, operations.

2. **Cause Resolution.** Focused on determining the cause of the failure so that a permanent

solution can be identified and, if approved, implemented.

Both of these problem-solving techniques are important and complementary. When they are properly applied, they provide value to the organization by maintaining/increasing asset/plant performance. Asset performance is usually referred to as the *Overall Equipment Effectiveness (OEE)* of the asset. OEE is a measure of how much good product we produce compared to what we could have produced and is a product of availability, machine speed, and quality yield. Failures introduce downtime, which impacts availability and, in turn, OEE. Solving the issue rapidly and/or preventing its recurrence have a direct impact on the asset's performance and the site's success.

In this book, we will dedicate a chapter on troubleshooting techniques and the steps required to be effective and consistent when troubleshooting problems. As the more involved techniques are focused on cause recognition and resolution, the majority of the book will be focused on *Root Cause Failure Analysis (RCFA)*. There are many tools available to conduct proper RCFAs, and it is important to understand what they are and when to use them. Too often we define RCFA as an Ishikawa or a 5-Why exercise and do not consider the full breadth of techniques and steps available and required to achieve and sustain the desired outcome. As a result, too many RCFAs are done but never implemented or deployed.

This book will detail the practical methods and tools needed to achieve our problem-solving goals and objectives, both when troubleshooting a problem and when determining a permanent solution.

Quantifying the Benefits

Asset performance improvement initiatives that are based on an increase in physical asset reliability are an excellent way to maximize financial return from our assets. These initiatives provide significant and sustainable benefits for relatively low financial investments, as compared to their capital expenditure alternatives. Problem solving, which includes troubleshooting and RCFA, is an important contributor to increased physical asset reliability. Beyond the value proposition of high return for low cost, reliability improvement programs drive the culture change that is needed to ensure sustainability because they include not only physical improvements to our asset base but also an improvement on how we manage our business and the systems and processes that support our business. Problem-solving techniques provide a continuous improvement stream that leads to an increase in benefits beyond the immediate incident being addressed.

As with all projects and initiatives, problem-solving activities need to be managed and, therefore, measured. The measures and expected results should be defined up front and used to scope out the activities and track success. Problem-solving strategies can help increase asset performance and asset life, reduce safety and environmental incidents, and reduce costs.

Some organizations qualify and quantify the cost of future failures prevented by RCFA activities as risk avoidance. To do this, we need to quantify the value of this occurrence (i.e., the number of safety incidents, downtime, or cost of the repair) and project the number of similar occurrences per year based on historical data. This is a good exercise, but it takes a lot of effort to develop, maintain, and even defend the results. That said, information of this kind helps foster support for an RCFA program within our organization.

We too often think of the cost of the failure as downtime or quality losses. However, the costs can be much more than this and other costs can outweigh the downtime and quality losses. Documenting and quantifying these costs are important parts of the RCFA analysis because they will help us to define the "value added" from the effort expended to analyze the problem. For example:

Compensation and benefits costs:

- Medical and rehabilitation costs for the victims and even for those not directly involved (i.e., mental trauma for the first responders)
- Pension and lump sum payments if the person is unable to return to work
- Death benefits
- Short- and long-term disability costs

Legal and legislative costs:

- Legal fees to defend this in court and/or for documentation and advice
- Penalties, fines, and citations; typically, government-applied (e.g., an environmental incident)
- Expert witness(es)
- Settlements
- Union grievances (these take people away from their daily activities and can sometimes result in settlements)
- Public relationships

Quality, productivity, and resource costs:

- Product replacement for damaged or substandard product, as well as disposal of the rejected product and/or the returned product (including shipping and handling costs)

- Time to observe the accident, accompany the victim to a hospital, and file reports
- Investigator's time
- Expert time and fees
- Cleanup time and material
- Repair of equipment and/or facilities
- Hiring and training replacement workers, including lower productivity because they are new
- Overtime to repair the equipment and/or increase operational time to compensate for the incurred downtime
- Reassignment of personnel to other roles, including light-duty activities
- Transportation costs
- Ramp-up time for the machine, post-failure

Other costs:
- Loss of customers and/or customer confidence
- Replacement of damaged parts, components, or assets
- Rental costs for alternative machines, lifting equipment, electrical generation equipment, etc.
- Emergency supplies

When embarking on an initiative like RCFA, it is necessary that we quantify the potential benefits and establish, measure, and communicate these success metrics. Key Performance Indicators (KPIs) might include:

- # RCFAs conducted/year
- % RCFAs implemented <3 months
- % RCFAs that met their objectives
- Cumulative value of the completed RCFA (cost avoidance)
- *Mean time between failure (MTBF)*
- Total downtime due to a failure (usually monthly downtime)

Problem Solving and Leadership

Establishing problem-solving processes, tools, and competencies is an essential element of a successful asset management-focused organization. But this does not happen on its own, and it requires strong leadership from management and from those tasked with leading the troubleshooting and RCFA activities.

A leader supports, encourages, coaches, and empowers others to perform in a team environment for the betterment of all and to meet the goals set out for the activity. He/she strives for an effective team culture and good communication. The leader is adept at resolving conflicts and at motivating others.

An Effective Team

To be successful at problem solving, organizations need to establish effective teams. This is a group of people who, as a whole, possess the skills needed to accomplish their function within the organization. Team members can work independently while sharing responsibilities. They share a common goal and are accountable for the team's performance. Members of an effective team consistently demonstrate a high level of collaboration and innovation, and they pursue a common goal.

Leadership is not vested in a single individual. Rather, various team members take it up, according to need. Leadership is recognized as everyone's responsibility. This results in a high level of focus and energy as well as mutual trust.

A leader not only leads the team but also helps the members become effective leaders and team players. Management is concerned with doing the right thing, and leadership is concerned with doing things right. A good manager/leader is concerned with doing the right things the right way.

The leader provides guidance and direction to people and is accountable for the success of the activities. He/she delivers results and does so with a collection of people who have varying objectives and personalities.

Being a leader is not an easy task. The leader is an example for the team members, follows processes, and communicates. A leader also has to deal with conflict resolution and keep the team motivated.

Communication

When communicating, the leader has to ask him/herself:

- Am I sending the right message?
- Am I using the right vehicle to communicate that message?
- Is the message understood and applied?
- Are we creating the right culture?

A leader must direct, delegate, and assign work to his/her team members (or accept work given to him/her), but that is only a small portion of his/her responsibilities. The way that directions are communicated is key to the results. They will travel across groups and within the organization and must mesh well with other messages. Every message, direction, request, or order must be aligned with the organization's values, mission, and objectives. This includes the non-verbal communication concerning how the work should be accomplished, the required behavior and attitude towards the activity, the rules that apply, and our commitment to safety and society as a whole.

Communication creates culture and ensures quality work and adherence to goals. When information is misinterpreted or disregarded, this is seen by other groups and at other levels in the organization. Poor communication within your group will affect its ability to work as a group and to work with other groups. Poor communication causes confusion and frustration. It can impact an individual's—and ultimately the group's—performance.

How to communicate as a leader:

- Listen to your manager and team members.
- Ensure that your behavior and actions are aligned with the company's values.
- Do not feel superior to your team members. Rather, be empathic.
- Do not leave holes in your messages. It does not matter if you have to explain it again and again.
- Tolerate criticism. Criticism indicates that you are not sending the right message.
- Do not hesitate in making decisions.
- Enjoy your work!

Conflict Resolution

On a daily basis, people interact with people who have varying personalities. This is especially true when trying to resolve a problem. Issues related to how people work and the need to meet a schedule are often the topics of gossip and criticism, which in turn create interpersonal conflicts. The same is true when conducting a trouble-shooting exercise or during an RCFA. In both cases, the activity is required because there is a problem (someone was hurt, there is an environmental compliance issue, production is impacted, there are high quality losses, etc.), and it is not unusual for it to lead to tension and frustration. This in turn can lead to conflicts.

A good leader must know how to resolve these conflicts within the team. He/she recognizes and responds to important matters, is ready to forgive and forget, seeks compromise, avoids punishing, and believes that a resolution can support the interests and needs of all parties. Think of a situation where the production line is down and a team of mechanics is tasked with getting the equipment back up and running. The clock is ticking, and the company is losing money. The production manager is staying close to the action and is getting regular calls from the plant manager—and sometimes from corporate—to find out what is taking so long. Tempers can flare up, leading to mistakes and possibly more downtime. Does this sound familiar? A good leader has to quickly understand both sides of the possible conflict and create an environment where the organization's needs are met (line back up and running) as quickly as possible, in a safe manner, and with the correct solution.

A good leader should be able to identify the conflict before it becomes an issue. The leader must not ignore matters of great importance to the other person; have explosive, angry, hurt, and resentful reactions; ignore the people involved, resulting in rejection, isolation, shaming and fear of abandonment; or expect the worst. All of these will exacerbate the situation.

The leader can apply the following four conflict resolution techniques:

1. Quickly relieve stress
   - Remain relaxed and focused in tense situations.
   - Remember that you control the situation.
2. Be empathic
   - Try understanding the problem without compromising the outcome.
   - Listen and try to understand the other person's perspective.

3. Improve your nonverbal communication skills
    - Important information exchanged during conflicts and arguments are often communicated nonverbally.
    - Nonverbal communication includes eye contact, facial expression, tone of voice, posture, touch, and gestures.
4. Use humor and play to deal with challenges
    - You can avoid many confrontations and resolve arguments and disagreements by communicating in a playful or humorous way.
    - Sometimes the conflict is related to the amount of stress and not so much about the problem. Being playful can make each party realize that the conflict is not that important.

Motivation

Motivation is important, as an unmotivated person will not do a good job and may be constantly involved in conflicts. Motivation will directly impact the team's ability to solve the problem, be it during a troubleshooting exercise or during an RCFA.

How do leaders keep team members motivated? A leader knows what the team needs, each member's expectations, and how to address these to get the most performance out of the team. A leader keeps his/her team constantly motivated by giving importance to each task on the business and personal levels and challenging and supporting each member. These leaders focus on the team members' potential and challenge it so that they grow personally and professionally.

1. Communicate constantly
   - By letting them know everything that is going on in the analysis, team members feel that they are an important part of the team.
2. Be tolerant and be a teacher
   - Do not expect everyone to do their job perfectly, and do not get frustrated if they do not know how to do it. Instead, teach them how to do it, and tell them that they have the skills and knowledge to get the job done.
3. Be positive
   - Always keep a positive attitude, even if things are not going well. Remember that other team members see you as an example, and they will try to imitate your attitude. A positive attitude motivates people.

4. Manage frustrations
    ▪ Jobs can get stressful. If frustrations become too high, let the people take a break—even if the line is down. It is amazing how a short break can make all the difference in identifying and applying the solution.
5. Be a friend
    ▪ Listen to them. Your team members see you as a leader and as an example. This will make them look to you for advice about their personal lives.

Chapter and Section Overview

This book is divided into three parts: Part I: Understanding Failures; Part II: Troubleshooting; and Part III: Root Cause Failure Analysis (RCFA). We chose to include both troubleshooting and RCFA in the same book, as a successful problem-solving strategy must incorporate both short-term strategies to get the assets back up and running quickly (troubleshooting) and the longer-term ones where permanent solutions are identified and applied (RCFA).

Part I: Before discussing how to resolve a failure we must define what is a failure. Chapter 2: Understanding

Failures reviews our understanding of failures, what they are, their causes, and types of failures. This chapter introduces us to failure modes and effects and what rules to follow when faced with a failure. This is the starting point of any problem-solving exercise: Before we can solve a problem, we must understand and recognize it.

Part II: Chapter 3 focuses on the troubleshooting process from when to use it, asking structured questions, the process itself, the qualities of a good troubleshooter, and what to do after the troubleshooting activity. Troubleshooting is generally the first process used when faced with a failure, because this will eliminate (or at the very least mitigate) the consequence of the failure until we can properly address its root cause. Therefore, good troubleshooting is a key element in managing our physical assets' performance.

Part III of this book puts into practice the concepts explored in the previous sections. It includes two additional chapters: Chapter 4: Overview of RCFA and Chapter 5: Conducting an RCFA.

Chapter 4 focuses on the RCFA methodology and explains how it differs from RCM, FMEA, FMECA, and PMO. It also focuses on when to apply RCFA. We discuss the cost of incidents or failures and establishing parameters as trigger points for launching an RCFA. The chapter introduces us to various techniques to collect and organize failure information and techniques to analyze this information in order to identify the root cause and a solution to resolve it. We will see how these techniques differ and when to use one versus the other.

Chapter 5 focuses on applying the RCFA techniques described in the previous chapter using a structured framework called the "8 Disciplines." We discuss the role of the facilitator in conducting an RCFA and the skills of a good facilitator. This chapter is about how to achieve consistent results in applying the RCFA process and how to make RCFA part of your company culture.

PART I

UNDERSTAND
FAILURES

CHAPTER 2

Understanding Failures

When we speak of failures from a maintenance and reliability point of view, we refer to the equipment's inability to meet its intended function (what we want it to do). This is called a *functional failure*. For protective devices whose failure may not be evident under normal circumstances, the functional failure is defined as the device not meeting its intended function when called upon to do so.

A functional failure might represent a total failure. It might also be a partial failure; that is, the equipment continues to function but it performs below the desired or required level of performance. These are more difficult to identify because most organizations are poor at setting clear performance expectations for their assets beyond expectations at the production line or plant level (i.e., so many tons per day shipped).

All functional failures are caused by something. The cause of failure is called a *failure mode*, and how it manifests itself is called the *failure effects*, which lead to the equipment's functional failure. When solving a failure, we want to solve the failure mode that caused the failure rather than the effects that were observed. Otherwise, the failure will repeat itself.

The failure might entail materials, products, structures, or components that do not operate or function as intended. This can cause personal injury, impact on the environment, damage, or economic loss. The purpose of a problem-solving activity is to locate cause or causes of failure with a view to improving the performance or life of a component and/or mitigating its consequences, either to resume operations and/or to prevent a reoccurrence of the failure.

The starting point is the failure itself, and being able to properly diagnose the problem is a key factor in solving the issue and helping the organization meet production demands, safely—while respecting the environment—at the lowest cost.

This chapter will cover:

- Failure Modes and Effects
- Causes of Failure
- Types of Failures
- Rules to Follow When Faced with a Failure

Failure Modes and Effects

Failure modes and effects are integral to Reliability Centered Maintenance (RCM), Failure Mode and Effects Analysis (FMEA) and Root Cause Failure Analysis (RCFA). "Failure mode" is a term used to describe any event that causes a failed state, listed at an appropriate level of detail. These descriptions consist of a noun and a verb; for example, "bearing worn" or "cable damaged." Failure modes can also be called "Causes of Failure."

"Failure Effects" describe what happens when a failure occurs and nothing is done to prevent it. The effects provide enough details so that we can choose the right strategy to mitigate the failure consequences.

Failure Modes

In practice, we stop listing failure modes once we have defined those that are reasonably likely to occur. This includes failure modes that have occurred before on the same equipment, are presently being prevented by existing PM/PdM programs, or have not occurred yet but are thought to be real possibilities. Less likely failure modes should be listed if the consequences are serious (e.g., health, safety, or environmental issues. We should also include failure modes that have a significant impact on product quality, downtime, and costs).

Example 1

A paper manufacturing plant has a subsystem that pumps 500 GPM of pulp stock from the blend chest to the paper machine's headbox. The reasonably likely failures (failure modes) that the system might experience include:

- Suction inlet blocked with a foreign object
- Pump drive end bearing insufficiently lubricated
- Breaker fails open
- Driveshaft key sheared
- Suction valve left in a closed position
- Impeller worn
- Bearing worn
- Packing worn
- Water ingress in the motor's junction box

Example 2

The railroad industry uses sophisticated circuitry to detect the presence and location of trains on the track in order to prevent collisions and optimize their movements. It is imperative for the connecting cables and their circuits to be in good condition. In this case, reasonably likely failure modes can include:

- Ballast contaminated with silt and dirt
- Insulated signal joint deteriorated
- Track circuit conductor fatigued
- Snowblower duct shifted out of alignment

Failure Effects

In the RCM and FMEA environment, failure effects describe what would happen if nothing was done to prevent the failure from occurring. It is important that we record adequate information describing the effects in order to be well enough informed so we can make the right decisions. In the typical RCFA environment, the failure effect is what we are trying to prevent. Therefore, the effects include not only the effects but also the failure itself (which we call a functional failure). In the Single Functional Failure RCM method (which we will discuss in Chapter 4), we distinguish between functional failures and failure effects as done in RCM and FMEA.

The following are examples of failure effects:

- With use, the v-belts wear, crack, stretch, start to slip, and, eventually, a belt breaks. The other belts are subjected to additional load and will eventually break as well. The compressor pump stops turning, and the loss of compressed air shuts down the process. Downtime: 1 hour. Cost: $2,000 US.

- The pressure vessel's pressure relief valve fails under normal conditions. In the event of an over pressure condition, the pressure in the vessel increases and can rupture. There might be people in the vicinity.

- With use, the bearing wears and generates noise (squeaking) and heat. Eventually the bearing seizes, can score the shaft, and the motor trips on overload. The process shuts down. Downtime: five hours. Repair costs to change bearing: $1,000. Repair costs if motor and/or shaft are changed: $10,000.

- With exposure to the environment, the railroad signal conductors age, which results in the loss of conductivity. This failure is intermittent, and there will be a voltage drop when it occurs. Loss of insulation will not have an immediate impact on operations. The increased resistance in the circuit causes the relay to drop out, falsely indicating the presence of a train. In falsely detecting a train, other trains are restricted from entering this section of rail, unless directed to under restricted speed.

Causes of Failure

There are three categories of causes of failures: physical, human, and *latent* (i.e., system/process). When analyzing a failure, we normally consider the physical causes first, then the human causes, followed by the system causes.

Physical Causes

Physical causes are usually some type of mechanical or electrical failure. It might not be the broken component itself but rather what caused it, such as mechanical fatigue or a specific type of corrosion from a particular source. It might be upstream or downstream from where the problem has manifested.

Physical causes are generally the most observable ones because they are accompanied by physical evidence. When considering physical causes, we look at signs, such as noise, heat, color changes, alarms, and equipment performance. For protective devices whose failure is not evident during normal operations, we can test the equipment to see if it is capable of performing its intended function if so required. For performance degradation observations, it is important to quantify what minimum performance is required of the asset so that we can compare current state to this standard.

Human Causes

Human error is a deviation from intention, expectation, or desirability. Human error can include:

- Wrong response
- Lack of response
- Failure to follow or comply with established guidelines or procedures
- Shortcuts in doing a job
- Communication deficiencies between crews or shifts

Human error is a deviation from intention, expectation, or desirability. Logically, human actions can fail to achieve their goal in two different ways: the actions can go as planned, but the plan can be inadequate (leading to mistakes); or, the plan can be satisfactory, but the performance can be deficient (leading to slips or lapses). A mere failure is not an error if there had been no plan to accomplish something in particular.

Human errors can be grouped into four categories: anthropometric factors, human sensory factors, physiological factors, and psychological factors (from John Moubray's book, *RCM II*):

- Anthropometric factors involve physical size and space. These are errors that occur because a person (or part of a person, such as a hand or arm) simply cannot fit into the space available to do something, cannot reach something, or is not strong enough to lift or move something.
- Human sensory factors involve a person's sensory limitations. For example, these errors might occur because a person cannot see (their field of view or color schemes) or cannot hear (background noise levels).
- Physiological factors involve working conditions. These errors occur because of environmental stressors outside of the worker that can reduce their performance, such as temperature, vibration, fatigue, or humidity.
- Psychological factors, on the other hand, involve actions taken by the employee his or herself. These factors might be intentional or unintentional.

Possible failure modes for human errors are errors made while operating or maintaining the equipment or during its design, fabrication, or installation. If human errors can be prevented, we need to include these in our problem-solving analysis. Prevention can include communication processes, training programs, documentation practices, visual aids, redesign, or operating procedures, PMs, installation practices, or work instructions.

When considering human causes, the investigation should be a search for the facts rather than a search for the "guilty." The objective is to get the situation into the open so that we can work to correct deficiencies and prevent the error from recurring.

Latent Causes

Latent causes are system- or process-related. The person might have made a human error, but we need to investigate whether it was caused by something outside of his or her control. Behind most human errors is a flawed system or a cultural aspect of the corporation or facility. Lack of—or deficiencies in—training programs, operating procedures, rules, policies, and guidelines contribute to human errors, which then allow physical failures to occur. Uncovering and modifying management system problems will not only prevent a single failure from happening again, but also possibly a series of others.

The search for root cause should never stop with an individual because of a human error; rather we should follow it further to determine why that person made the error. Poor procedures, inadequate training, inadequate specifications, or facility culture may be the system root cause, and any number of personnel in the same position at the same time might have made the same decisions that led to the failure. For example, if a mechanic replaces the wrong valve, we usually categorize this as a "human error." However, in order to determine the *root cause* of this failure, we need to go further and ask:

- Was the procedure clearly defined?
- Was the valve properly tagged?
- Is the mechanic familiar with the task and work area?

When solving latent causes, we need to determine what improvements in systems and processes would prevent the problem from recurring. Further, we need to ensure that corrective actions remain in place and are successful.

Types of Failures

There are two types of failures: sporadic and chronic.

Sporadic Failures

Sporadic failures usually happen infrequently, suddenly, and dramatically, and lead to significant consequences. We say "infrequently" because if the company or site had sporadic failures that occurred frequently, it would not stay in business. The failure, clean up, and problem resolution require the use of many resources, and the activities are very visible, drawing significant organizational attention. Sporadic failures include explosions, fires, major leaks, and safety and environmental incidents. Generally, we solve these failures as an emergency and continue with our normal activities once the issue is resolved.

Sporadic failures are usually caused by one failure mode, though the cause is not always well understood. Because their consequences are significant, and the failure is highly visible within the organization, it is not unusual for multiple actions to be taken to resolve the failure, as we observe it, so that production can resume immediately. As a result, we might have solved the effects but not the root cause of the problem (due to organizational pressure for immediate action), discarding parts, repairing equipment, and restarting the process with little investigation into the failure's cause.

Sporadic failure investigations are sometimes formal, and some of them are mandated by the *Occupational Safety and Health Act (OSHA)*. We should also use troubleshooting techniques to resolve the immediate issue and then apply an RCFA method of investigation to identify and permanently resolve the root cause.

Chronic Failures

Unlike sporadic failures, chronic failures are generally not dramatic or difficult to repair, and they usually occur gradually. However, it is more difficult to identify the root cause. These failures tend to be repetitive or short-term, affecting production and/or maintenance. They are also accepted as an operational cost, such as impeller worn, bearing seized, seal worn, or drive belt worn. We become so adept at resolving these that they become part of the status quo and taken into account when setting production quotas and maintenance budgets. Chronic failures tend not to get the attention of sporadic failures because each individual occurrence is usually not very costly, although, cumulatively, they might be. Thus, resolving them can have a positive impact on site performance and morale.

For example, a bottle washer for returned bottles on a bottling line was experiencing 40 jams per shift at its infeed due to fallen bottles. It takes 90 seconds on average to unjam the infeed. As a shift is eight hours and they run 24 hours/day, the line was losing three hours of production per day (12.5% unavailability). Initially, the problem (though known) was not quantified and usually attributed to the condition of the bottles. A closer investigation showed us that the jams were predominantly caused by the configuration of the infeed rails and conveyor and, to a certain extent, on the setting on the case bottom saws (used to remove the cases from the bottles). Both could be resolved.

A similar situation is a pump impeller failing because of gradual wear. As the impeller wears, the pump's output is reduced. We can measure this output by the reduced flow from the impeller or reduced electrical current draw to the motor. However, most operators adjust the setpoint and/or adapt the process if possible. We learn to live with (and gripe about) the reduced flow until the impact to production is too great or management raises the flag. This is then managed as emergency or break-in work.

Most organizations fail to consider the total impact of chronic failures. Since the chronic event is usually viewed as an individual occurrence (whose consequences are not too severe), it is easy to overlook the accumulated consequences and related costs (which can be very severe).

When we systematically identify and eliminate the root cause of chronic failures, the equipment, process, and plant outputs begin to rise because we are no longer limited by the accepted status quo (resulting from equipment degradation) and are able to utilize the assets at or near their design limits. Downtime and repair costs are reduced, as is the overall frustration among the production and maintenance teams.

Rules to Follow When Faced with a Failure

Failures are always stressful and create tension among stakeholders. While we do not want to be nonchalant about the failure, being focused on finding a solution is different from being stressed. One brings positive energy while the other brings negative energy that can lead to errors. In either case, we need to remain professional and avoid turning the problem-solving exercise into a witch hunt. We are not looking for the guilty party, but we are looking at how to mitigate the consequences, resolve the issue, and prevent its recurrence.

The four most important rules that we need to follow when we are faced with a failure:

Don't Panic

When a failure occurs, time constraints (and perceived time constraints) can cause us to panic and act too quickly, which often leads to crucial mistakes or taking unnecessary actions. This can be prevented by breaking the problem into its parts and tackling each one individually. In order to do so, we need to consider the symptoms that we observed, when the problem occurs (e.g., at start up, while stopped, or during operations), the desired state of the equipment, and what we know and do not know about the problem.

Keep an Open Mind

No matter how much we know about our equipment, we do not know everything. Because of this, we need to accept inputs from inside and outside our department as well as from outside sources and take the time to consider these opinions. This can help us to think outside the box when we cannot think of an immediate cause of, or solution to, a problem.

Stabilize the Situation

In order to stabilize the situation after a failure occurs, we consider the following issues in order: safety, the environment, and production. If the failure is a safety issue, we need to help anyone who was impacted and prevent anyone else from being injured. If it is an environmental issue, we need to contain the situation (e.g., containing a spill), take measures to resolve it, and inform the appropriate authorities. If production is affected, we need to consider the impact to throughput, quality, customer service, and cost. We also need to consider how quickly we can resolve the issue and if we need to apply contingencies.

Preserve Evidence

In order to understand the cause of the failure, we need information. This means that we need to resolve the immediate effects and preserve as much evidence as possible, as soon as possible. This includes interviewing those present, taking photos, recording process conditions, and bagging/tagging failed parts. We do not want the situation to reoccur, and, in order to avoid recurrence, we need to determine and implement the right permanent solution.

Use a Structured Approach

In the next chapter we will talk about the structured approach: symptom, cause, and remedy. This is a structured approach and the best troubleshooting one for most situations. It follows three steps: be structured, follow a process of elimination, and don't jump around. If we are structured, this will help others become structured and will help us to identify the failure and its solution faster and correctly.

———————

This chapter discussed understanding failures, focusing on failure modes and effects, causes and types of failures, and the rules to follow when faced with a failure. The next chapter focuses on the troubleshooting process: when to use it, levels of rigor, asking structured questions, the process itself, what to do post-troubleshooting, and the qualities of a good troubleshooter.

PART II

TROUBLESHOOTING

CHAPTER 3

Troubleshooting

According to the *Oxford English Dictionary*, "trouble-shooting" is "the process of analyzing and solving serious problems for a company or organization."

Unexpected failures happen. In a normal production facility, dealing with these failures is a daily occurrence for the maintenance and production departments. We can reduce the probability of failures by applying reliability methodologies (i.e., *Reliability-Centered Maintenance (RCM)*, *Failure Modes and Effects Analysis (FMEA)*, and *Life Cycle Costing (LCC)*) and workforce effectiveness practices (i.e., precision maintenance, *Operator Asset Care (OAC)*, and autonomous maintenance). How we deal with failures is an essential part of our plant's success, both in reducing unscheduled downtime and reducing maintenance costs. The first line of response to a failure is troubleshooting because this can be done immediately and helps return the asset to an operational state as soon as possible.

Troubleshooting is the most basic and practical diagnostic method. It allows us to quickly find a solution so that we can resolve the problem and continue operating the equipment without significant consequences. Troubleshooting might require that we try a number of actions in order to solve the issue, but troubleshooting does not focus on the root cause so that we can introduce a permanent solution. This is not to say that a permanent solution will not be found, but rather that this is not its main objective—getting the equipment up and running is. When troubleshooting, time is of the essence.

Understanding what we expect from our troubleshooting activities and aligning our organization accordingly will make employees more effective and efficient and help us to better define the role of subsequent methodologies, such as RCFA. In simple terms, when we troubleshoot a problem in a production facility, we want to minimize downtime and quality losses as usually every minute and percentage loss costs money. As in-depth analysis takes time, a luxury that we do not have at this moment. This will come later.

When we use troubleshooting techniques, the process we employ in leading the activity, what skills are required and what we do once the problem is resolved are all key to maximizing the value of our troubleshooting activities.

This chapter will cover:

- When to Use a Troubleshooting Process
- Levels of Rigor
- Structured Questioning
- The Troubleshooting Process
- Post-Troubleshooting
- The Qualities of a Good Troubleshooter

When to Use a Troubleshooting Process

An obvious answer to this is "when we have a failure." We should also use it when we notice a degradation in asset performance. On the other hand, we should not use a troubleshooting process as a means to identify the root cause of the problem or solve safety or environmental issues. Troubleshooting is a tool that we use to immediately remedy the situation so that production can resume at or near the desired level of performance. The solution may or may not be permanent.

Also, avoid using a troubleshooting process to resolve a known problem and solution. For example, the equipment is being operated beyond its design capacity. True enough, your people can do amazing things and may be able to find ways to keep the equipment running, even in this situation. But over time, this is a frustrating situation for all and may eventually impact motivation and work standards, where people patch things rather than properly repair them. Among other things, the site's troubleshooting practices will degrade.

We respond to unexpected failures by establishing strong troubleshooting practices and training our people accordingly. Technology can help us to foresee the eventual problem, but troubleshooting requires human reasoning. The results of poor troubleshooting are excessive downtime, high maintenance costs, higher number of safety and environmental incidents, and loss of control of our maintenance process. Therefore, troubleshooting techniques should be mastered by everyone in the organization and used on a daily basis.

Levels of Rigor

Not all failures and incidents require the same level of rigor to analyze, identify, and implement a strategy. The type and severity of the consequences of the failure are factors that will influence this. For example, to a certain extent, we will ignore time and effort when analyzing a safety, regulatory, or environmental incident. On the other hand, when analyzing a failure that caused the line to stop, we will not want the troubleshooting exercise to add to the downtime. The duration of the trouble-shooting activity becomes important, outweighed only by the severity of the failure. But for a downtime-related incident, we must take time to ensure that the solution (which might be temporary) will allow the line to continue to operate adequately until the next scheduled line stoppage, but that in no way will introduce a safety or environmental risk.

When considering other consequences, such as quality losses or excessive operating costs, we want to address the problem in such a way as to eliminate—or at least greatly reduce—the losses and additional costs. We should not underestimate these factors because they can be significant.

Unfortunately, we tend to gravitate towards the level of least effort. As a result, we too often do a poor job of troubleshooting and have to revisit the problem a number of times until it is fixed properly. When troubleshooting an incident, the production supervisor and maintenance technician (the maintenance supervisor is often involved as well) assesses and decides if the problem can be dealt with in quick fashion or if it demands greater input and resources to properly solve the dilemma. The time limit for this decision can be set by the facility or plant management, depending upon the production constraints. For example, in the automotive industry where every minute of downtime represents thousands of dollars of production loss, quick solutions are a must.

We should also consider the likelihood of a temporary solution providing the desired outcome until we have time to properly address the problem, the potential secondary damages of not properly resolving the issue, and the time/cost to do it right the first time compared to doing so in the future.

For example:

- A screw conveyor trips on overload. We reenergize the circuit breaker and restart the conveyor. Problem solved? We should ask ourselves, why did the conveyor trip? It could be a downstream issue (partial blockage), a piece of foreign matter in the conveyor, loose screw flights, or an operator error where the conveyor was fed too quickly. Even in the latter case, if the conveyor was fed too quickly, did this cause damage? Is the screw always centric, and is there any unusual noise? Will this blockage cause any downstream blockage? Not troubleshooting the failure past restarting the conveyor could result in continued issues and in more downtime than if we had properly diagnosed and resolved the issue the first time.

▪ A copper casting wheel stops because of a dirty photocell. If the wheel is stopped for more than five minutes, we need to cool it down, clean it, and reheat it. This takes eight hours. In this type of scenario, we should look at solutions that can be quickly (in less than five minutes) applied. We should probably try to restart the wheel, if the control system does not indicate the photocell as an issue. Next, we clean the photocell and try to restart the wheel. If neither solution works, we are down for eight hours and should therefore take the time to properly diagnose and resolve the problem.

In either of these cases, we need to ask ourselves: "Have I identified the cause of the failure to the level at which I can anticipate it and apply the appropriate solution?" and "Will the asset continue to operate adequately until the next scheduled stoppage or until we can properly address the problem?" If the answer is "no" to either question, you have not gone deep enough in your analysis. If the answer is "yes," your analysis is detailed enough.

Structured Questioning

When troubleshooting a problem, we are too often faced with inputs from many sources. Unfortunately, we receive many opinions and too little useful facts. Because the purpose of the troubleshooting exercise is to get the asset back up and running as quickly as possible without compromising safety or the environment, we need to identify the facts as quickly as we can. To do this, the facilitator uses a structured questioning approach, which can include open and closed questioning, and/or Socratic questioning.

Open and Closed Questioning

This type of questioning is simple and straightforward and can help us to solve a problem and get the equipment back up and running.

A closed question is generally one that can be answered with a "yes" or "no" (Were you home last night?) or a simple statement, such as: "50 psi." We use closed questions to guide the person to the right answer quickly and prevent deviation in the discussion.

We use open questions in order to probe further. They allow for the person answering to elaborate. "Why?" is a common type of open question. Others include "How does this work?" or "What happened?" Once the answers provide sufficient insight into the cause of the problem, the facilitator uses closed questions to narrow down the discussion and come to a resolution.

Left without guidance, many discussions will take much more time than needed and will not provide the desired results. At the same time, if you are too directive in your questions, people will stop talking and simply tell you what you want to hear; they will stop being invested in the conversation. A proper blend of open and closed questions provides an open forum so that people feel free and are encouraged to share their thoughts while structuring the discussion so that it provides meaningful results.

Socratic Questioning

Police, doctors, and lawyers are trained using Socratic questioning methods, which are designed to quickly guide people being interviewed to the facts. These take their origins from Socrates, an ancient Greek philosopher who was famous for drawing out answers from his students; this included Plato, who then taught Aristotle.

Socratic questioning is guided and directive. When we use it, it helps us to find any flaws in our interviewee's answers, and it can provide us with more complete answers. This detailed questioning can lead us to the answers that we need, but it is more challenging than simple open and closed questioning. That is, we need to know what kinds of questions we need to ask to get the answers that we need. Further, Socratic questioning can only help us to find the answers that we need if we are listening to the answers that our interviewees give us. If not, then we can easily mislead these people by guiding the questions towards the answers that we *want* to hear rather than the ones that we *need*.

With this said, if we understand these questions and their functions, then we can become more effective interviewers, and—by extension—troubleshooters. There are six types of Socratic questions (Table 3.1).

Socratic questioning techniques are not learned overnight, as they require practice to master and incorporate into your natural way of approaching a problem. If this is new to you, start with the open and closed questions and gradually introduce the various Socratic questioning categories listed in the guide. With practice, add more and more categories until they all come easily to you. Don't forget to revisit the list every so often to reinforce the concepts.

Table 3.1 Socratic Questioning Guide

Question Type	Purpose	Examples
Conceptual clarification	To understand the person's rationale, and to make them think about their responses	• What exactly does this mean? • Can you give me an example? • What is the problem that you are trying to solve?
Challenging assumptions	To make the person interrogate the assumptions backing their argument	• Can you elaborate? • What else could we assume? • Is that always the case?
Examining rationale, reasons, and evidence	To ensure that the person's argument is backed by sound logic/facts	• Why? • Can you show me? • What evidence is there that supports this?
Considering alternative viewpoints and perspectives	To help the person consider other viewpoints or look at the situation objectively	• What if we compared these two options? • Why is this better than option B? • Who would be affected and what would they think?
Considering implications and consequences	To help the person interviewed to understand the consequences of their argument	• What would/could happen? • What if you are wrong? • What does our experience tell us will happen?
Meta questions	Reflexivity; turning the question over to the person asking the question in order to regain control of the situation	• What do you mean? • Why do you think I asked this question? • What else might I ask?

The Troubleshooting Process

The troubleshooting process identifies the following three parameters (Figure 3.1):

1. **Symptom.** Evidence that the equipment has failed or is failing (some people call this the "problem")
2. **Cause.** What produces the symptom (i.e., the cause of the symptom)
3. **Remedy.** Solution to mitigate the impact of the failure or to solve the problem

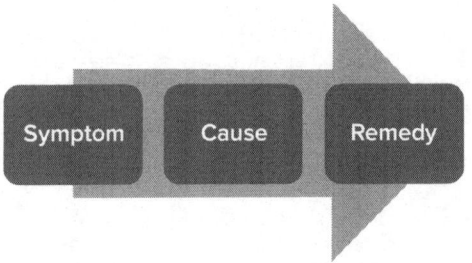

Figure 3.1 Symptom, cause, remedy approach to troubleshooting.

To answer these, we follow a systematic five-step approach:

1. Define the Problem (the symptoms)
2. Identify the Cause
3. Identify the Solution (the remedy)
4. Apply the Solution
5. Evaluate the Results

Step 1: Define the Problem (the symptoms)

The "problem" includes the symptoms that we and/or others have observed. This could be visual or audible signs, trends in performance, odors, temperature, vibration, performance of upstream or downstream equipment, or equipment setting changes. It should not be the conclusion (e.g., motor needs to be replaced), but rather what was observed and in which sequence it occurred (e.g., flow from the pump gradually dropped and the motor tripped). At this stage we want all the facts, as impartial as possible, so that we can quickly identify the right solution. We are not looking for the guilty party; we are gathering the facts only.

Some key questions are:

- What are the observed symptoms?
- When does the problem occur (i.e., at startup, while stopped, in operations, etc.)?
- What is the desired state of the equipment?

Step 2: Identify the Cause

In a troubleshooting activity, we may or may not be able to identify the actual root cause. This is because the focus of the activity is to resolve the issue as soon as possible and get the plant up and running. That said, we want to get as close as possible to the cause without delaying production.

For example, say that a conveyor stopped turning due to broken v-belts. We still need to look further to find the cause of that break. Was the load too great (i.e., overloading at the infeed)? Is the conveyor blocked? Are the belts worn or degraded? Are the pulleys out of alignment? If out of alignment, were they improperly aligned, or did they drift (i.e., broken taper lock)? If it is a broken taper lock, we will change it but will probably not have time to investigate why it broke. This analysis will have to be done outside of the troubleshooting exercise and might require an RCFA.

Although the 5-Why approach (see Chapter 4) is an RCFA-specific technique, we can and should still apply the 5-Why logic while troubleshooting. Therefore, we need to ask: "Why do we have these symptoms?" If the answer does not lead to the cause of the failure, then we need to ask "Why?" again until the cause (or causes) is uncovered.

The failure mode is the cause of failure at the level at which we can take action (in this case, immediate action) to prevent the failure or mitigate its consequences. In terms of consequences, we refer to:

- Safety
- Environment
- Quality
- Throughput
- Cost
- Customer Service

Step 3: Identify the Solution (the remedy)

In this step, we develop a list of possible solutions to the problem—independent of their value or practicality. Most people do this verbally, though writing things down always helps, and it can actually speed up the process. Through this brainstorming exercise, we might discover a solution that we had not previously considered. Therefore, it is important to not discount or accept a solution too quickly, though a certain level of realism is crucial. Once all of our ideas have been listed, we need to make a decision. The solution might incorporate more than one element. We need to evaluate the pros and cons of the solutions and follow a process of elimination.

Because the asset's performance is impacted by the failure and it is probably stopped, there is a lot of pressure to resolve the problem immediately. But as the adage goes, "Measure twice and cut once." The same rule applies to the brainstorming exercise. We do not want to waste time, but we do need to take enough time to properly evaluate the options and to make the right decision. In the end, this will save us time. This is often where the maintenance supervisor needs to intervene as he/she needs to make certain the troubleshooter is on the right track (and not wasting time). If so, they need to manage external pressures and expectations so as to allow the troubleshooter the time and space needed to resolve the issue.

Step 4: Apply the Solution

Once we have decided on the steps that we need to take in order to apply the solution, our next step is to *apply the solution*. If we have more than one solution, then we need to apply one solution at a time and take time to test it before applying the next one if the first did not deliver the anticipated results. We need to know exactly what resolved the issue, so that we can analyze the problem (at a later time) and find a permanent solution, if one is needed.

That said, nothing prevents us from setting up a backup solution (i.e., a "Plan B") while we test out our original plan. For example, we had an electrical drive on a critical equipment fail. The plant happened to be on a down day (it was stopped) but could not start up without it, and we did not have a spare drive in stock. We brought in a drive specialist to troubleshoot and hopefully repair the drive, searched the marketplace for a spare drive, and prepared a different drive so that it could be jury-rigged in place if the first two options did not work. Option A and B did not work, but our preparation to use the different drive allowed us to resolve the problem before startup.

From this example, we can see that we do not need to be absolutely certain of the solution before trying it; we can use a trial and error approach (within limits). At the same time, we still need to *think before acting*. Making a decision might lead to undesirable consequences, but it is often possible to evaluate the risks. Taking action brings satisfaction and self-confidence to the person who made the decision and to those affected by it.

Certain behaviors or fears might influence us against applying the selected solution. This is especially true if the company culture is one of blame. As an organization, we must acknowledge that people are doing their best and support their efforts, even those that fail. It is important to remember that, in any case, we should *never put safety (ours and others) or the environment at risk.* We should not execute unsafe jobs or jobs in an unsafe manner. If we have safety concerns, then we need to bring them up with management.

Step 5: Evaluate the Results

This is a very important step, and one that is often missed when we go from one emergency to the next. We need to give ourselves, or give our personnel, a bit of time to ensure that the problem is resolved. We need to assess whether we have met our objectives. If we have not, then we need to inform production and management. They might want to continue without a resolution, or they might direct us to start again. In the latter case, we need to repeat each problem-solving step. Certain elements might look different when we do so.

Once the problem is resolved, we should take a moment to see what we learned from the exercise. Were our initial thoughts correct? Did we follow the process? Should this solution be applied to other equipment? Was the process supported by good communication? What will we do differently next time?

Post-Troubleshooting

While troubleshooting fills the "Respond to the Incident" parts of the RCFA process (as outlined later in the book), it is important to emphasize the "Preserve and Record Evidence" element (again outlined later in the book) as well, so that we can develop a strategy to permanently resolve the issue, if so desired.

Therefore, if the failure had significant consequences post-troubleshooting, then we need to continue the analysis through a Kaizen, Blitz, RCFA, or simply meet to determine if the problem was actually resolved and, if not, to resolve it and define measures to prevent it from happening again. We might also want to monitor and track the situation for a period of time. Unfortunately, this is rarely done and too often our answer is to add something to our maintenance program or operating practices as a knee-jerk reaction, without applying a formal methodology. This is how our PMs and SOPs get out of control and how our people lose confidence in them.

It is also important to ensure that all documentation is up to date (including comments on the work order [WO]) and that people are informed and trained as needed so that knowledge is shared. We do not necessarily need to update the PMs and SOPs, but we simply ask that you do so in a formal review, so that these are optimized for your business needs and operating context.

The Qualities of a Good Troubleshooter

In a typical organization, the equipment breaks down and—though many people voice their opinion—it is not unusual to find out that no one really knows why it did. We talk to the operator and/or start opening the equipment. Some people might look at manuals and notes, but the information is usually not properly documented or classified. We try a couple of things, and the situation might be resolved or, at least, the effects seem reduced. Hopefully, our top mechanic or electrician is available and has seen this before. Chances are that the problem is not resolved. Unfortunately, before we can continue, another more important emergency pops up, and key resources are pulled away. The problem-solving exercise drags on because, as an organization, we are not structured to support the troubleshooting activity, do not properly recognize its value, and do not adequately train our troubleshooters.

Good troubleshooting starts with the company culture. As an organization, we must recognize the importance of troubleshooting and support the person in doing his or her job. This starts with recognizing that troubleshooting requires a structured approach for it to be successful and providing the troubleshooter the tools and time to do the job correctly. It also includes a firm understanding of the objectives of a troubleshooting exercise versus an RCFA, and setting the right expectations. Finally, the company must define troubleshooting as a practice and the troubleshooter as a role. This includes training specific resources, measuring results, and managing the process. It also includes hiring and/or assigning people with the right skills to this role.

A good troubleshooter:

- Is safety-minded
- Understands and can apply the five troubleshooting steps listed earlier in this chapter
- Listens attentively and asks meaningful questions
- Communicates effectively and keeps all stakeholders informed
- Thinks systematically and asks structured questions
- Keeps an open mind and can think outside the box
- Is confident and has self-esteem, does not panic
- Is honest and has integrity

- Is driven to get the job done
- Takes a positive approach
- Behaves professionally and is respectful of others

―――――

This chapter focused on the troubleshooting process from when to use it, the process itself, asking structured questions, the qualities of a good troubleshooter, and what to do post-troubleshooting. The next chapter will introduce us to *Root Cause Failure Analysis (RCFA)*, compare it to other work identification methodologies, identify when to use it, and help us quantify the cost of a failure. We will look at the different methods used to collect and organize the information in preparation for the eventual analysis and review three RCFA analysis methods.

PART III

ROOT CAUSE FAILURE ANALYSIS (RCFA)

CHAPTER 4

Overview of RCFA

Root Cause Failure Analysis (RCFA) is a method that we use to determine the cause of a failure once it has occurred at the level at which we can anticipate and take appropriate action to prevent future failures or mitigate their consequences. (The term *Root Cause Analysis [RCA]* is sometimes used interchangeably with RCFA; in this book, we will be using RCFA.)

RCFA is generally used as a continuous improvement methodology for analyzing and resolving a failure with high consequences whose cause is unknown, that has not been resolved through a formal approach (e.g., RCM or FMEA), and that is not being addressed through a modernization or CapEx initiative. RCFA is frequently applied to failures with safety or environmental consequences, but should only use the following techniques if they are not in conflict with your organization's safety or environmental compliance procedures.

What distinguishes RCFA from other problem-solving methods is that it incorporates detailed information collection, organization, and analysis stages, which are applied after the failure has occurred. Too often we find that organizations skip some of the vital stages of the RCFA, resulting in less than optimal outcomes. This might account, in part, for the high percentage of RCFAs that are conducted but not implemented. In this chapter, we will explain in detail the various techniques available to collect, organize, and analyze the failure information, as well as when to use one technique versus another.

This chapter will summarize the RCFA methodology and explain how it differs from RCM, FMEA, FMECA, and PMO. It is important to note that many of these methodologies use specific words in different ways, depending on the version that we employ. We have standardized on the RCMII terminology (John Moubray's book, *RCM II*) for all of the above, including our RCFA method (with the exception of the terms "functional failure" and "failure effect"). This greatly reduces the need to translate between methodologies and to train people on each method and its terms.

This chapter will also introduce us to the different methods that we can use to collect and organize the information in preparation for the eventual analysis. These methods are categorized as: Interviews and Reports, Data Collection and Classification, and Brainstorming Exercises. Finally, we will review three RCFA analysis methods, including the "5-Whys," "logic trees," and "Single Functional Failure RCM (SFF-RCM)."

This chapter will cover:

- RCFA Versus RCM/FMEA/FMECA/PMO
- When to Apply RCFA
- Collecting and Organizing the Information
- Analyzing the Failure

RCFA Versus RCM/FMEA/FMECA/PMO

As mentioned previously, RCFA is used post-failure and includes detailed failure information collection, organization, and analysis stages. RCFA is a methodology that analyzes a *specific failure* with significant consequences and identifies a strategy to mitigate its consequences if it were to happen again.

Reliability Centered Maintenance (RCM), Failure Mode and Effect Analysis (FMEA), Failure Mode, Effect and Criticality Analysis (FMECA), and Preventive Maintenance Optimization (PMO) are methodologies used to develop an asset care program that anticipates a large number of reasonably likely failure modes and identifies strategies to mitigate their consequences *before they occur*. Some methods are best suited for reviewing existing programs, while others are suited for developing new programs.

One way to depict the differences between RCFA and RCM/FMEA/FMECA/PMO is depth versus breadth. RCFA looks at only one failure incident and dives deep to identify a permanent solution, while the other methods look at many potential failures (i.e., they take a broad approach) and do not go to the same level of detail.

RCM is a methodology used to determine what must be done to ensure that any physical asset continues to do what its users want it to do in its present operating context. FMEA is a subset of RCM, while FMECA includes a criticality component to the FMEA process, and PMO follows a similar approach with an emphasis on optimizing existing programs. The latter three methodologies typically do not provide the same level of rigor as RCM, but are faster. All methods are facilitated and are applied at the system level, reviewing a relatively large number of reasonably likely failures and defining strategies to mitigate their impact on the organization.

We use FMEA and PMO over RCM (depending on the version of FMECA, this might require a similar level of effort to RCM) when resource constraints prevent us from applying RCM to all significant assets in a timely fashion. We also choose this route when the responsible asset managers must address their asset care program shortcomings within a constrained time frame and with limited resources.

With RCM, PMO, FMECA, and FMEA, only a small number of failure modes merit the level of failure information collection, organization, or analysis used in RCFA and usually, as the failure has not occurred, we do not have access to the same level of detail. Because we have to wait for the failure to occur, RCFA is a reactive way to become proactive. It is also focused on a very specific issue. Therefore, RCFA cannot and should not replace RCM/FMEA/FMECA/PMO; nor should they replace RCFA.

Because the consequences are significant, and we have the failure to investigate, RCFA usually includes an in-depth analytic element and can represent several days of work to solve this single failure. The main reason why we analyze a specific functional failure is to ensure that the failure and its associated consequences do not occur again. The consequences associated with any failure can fall into one of the following consequence categories:

■ **Hidden Consequences.** A multiple failure results from the failure of a protected function and its associated protective device

- ▪ **Safety.** The failure could result in injury or fatality
- ▪ **Environmental Consequences.** The failure could breach an environmental standard or regulation
- ▪ **Operational Consequences.** The failure could affect operational capability (throughput, product quality, customer service, and/or operating costs)
- ▪ **Non-Operational Consequences.** The failure involves only the cost of repair

Conducting an RCFA requires a team effort. The team members who need to be involved are an RCFA-trained facilitator, production, and maintenance experts, supervisors, and specific specialists or departments, as needed.

Organizations often only require RCFA analyses for failures that cause extended production unavailability. However, we should consider conducting RCFAs for failures that affect safety, the environment, or cause high repair costs, quality loss, production loss, or excessive downtime. We also cannot forget legislated requirements, such as environmental regulations. Our tolerance to some of the categories is dependent on the business unit and the impact a failure has on the organization. To this end, we might want to develop a decision matrix that specifies the minimal trigger points for each category (as seen later in this chapter). In addition to these categories, we should consider RCFA for failures that are chronic or repetitive, and that have a total impact over a specific period of time that is not acceptable or tolerable.

When to Apply RCFA

Oftentimes, organizations establish a threshold of two or four hours of downtime for when we can conduct an RCFA analysis. This does not remove the first step of conducting a troubleshooting activity to bring the line or plant back up but, rather, as a subsequent step to eliminate future occurrences of this failure.

But what about a failure that poses significant safety or environmental risks? Should we also consider the impact on quality and cost of the failure in our selection criteria? How about chronic failures (as discussed in Chapter 2)? When discussing chronic failures, we presented a case where the bottling line was losing three hours of production per day, at 90-second increments. Ninety seconds is well below the two- or four-hour downtime threshold, though we were losing three hours of production per day. Finally, even if the failure meets one of the selected thresholds, is RCFA always the best strategy? (We already know that this is a design flaw or the equipment is being used beyond its design limits.)

In order to be effective, the RCFA selection criteria must include multiple parameters, as shown in the following example from a steel manufacturing plant (Table 4.1).

As the earlier example shows, the steel plant chose to adapt the thresholds to the specific business unit. This is a great idea because not all business units impact the business the same way.

Once a failure has been identified as meriting the effort necessary to conduct an RCFA (and implement the results), we still have to decide if we will conduct one or not. The percentage of RCFAs conducted and not implemented is quite high (this varies from site to site, and, in some organizations, more than 80% are not implemented), so we must be selective. At one of our clients, management mandated that all failures causing a measurable production loss needed to be subjected to an RCFA. This resulted in a deluge of RCFAs, and very few of the results were ever implemented.

Table 4.1 Example of an RCFA Minimum Trigger Point Decision Matrix

Asset Type	Safety	Environment	Repair Cost	Quality	Production Loss	Downtime	Other
Blast furnace	1 or more LTI	Notice of violation	Greater than $300,000 in equipment breakdown	More than 75,000 tons poor quality	Greater than $300,000 in lost opportunity	Greater than 10 hours of consecutive unplanned downtime	At request of senior management discretion
Coke production	1 or more LTI	Notice of violation	Greater than $300,000 in equipment breakdown	More than 1 heat poor quality	Greater than $300,000 in lost opportunity	Greater than 8 hours of consecutive unplanned downtime	At request of senior management discretion
Sinter plant	1 or more LTI	Notice of violation	Greater than $100,000 in equipment breakdown	More than 1 heat poor quality	Greater than $100,000 in lost opportunity	Greater than 8 hours of consecutive unplanned downtime	At request of senior management discretion
Steel making	1 or more LTI	Notice of violation	Greater than $275,000 in equipment breakdown	Moe than 1 heat poor quality	Greater than $500,000 in lost opportunity	Greater than 8 hours of consecutive unplanned downtime	At request of senior management discretion

(continued on next page)

Table 4.1 Example of an RCFA Minimum Trigger Point Decision Matrix (*continued*)

Asset Type	Safety	Environment	Repair Cost	Quality	Production Loss	Downtime	Other
Caster	1 or more LTI	Notice of violation	Greater than $250,000 in equipment breakdown	More than 12,000 tons poor quality	Greater than $500,000 in lost opportunity	Greater than 4 hours of consecutive unplanned downtime	At request of senior management discretion
Hot mill	1 or more LTI	Notice of violation	Greater than $250,000 in equipment breakdown	More than 50 coils of the same consecutive defect	Greater than $500,000 in lost opportunity	Greater than 24 hours of consecutive unplanned downtime	At request of senior management discretion
Plate mill	1 or more LTI	Notice of violation	Greater than $250,000 in equipment breakdown	More than 50 coils of the same consecutive defect	Greater than $100,000 in lost opportunity	Greater than 24 hours of consecutive unplanned downtime	At request of senior management discretion
Pickle line	1 or more LTI	Notice of violation	Greater than $100,000 in equipment breakdown	More than 15 coils of the same consecutive defect	Greater than $100,000 in lost opportunity	Greater than 12 hours of consecutive unplanned downtime	At request of senior management discretion

(continued on next page)

Table 4.1 Example of an RCFA Minimum Trigger Point Decision Matrix (*continued*)

Asset Type	Safety	Environment	Repair Cost	Quality	Production Loss	Downtime	Other
Cold mill	1 or more LTI	Notice of violation	Greater than $100,000 in equipment breakdown	More than 20 coils of the same consecutive defect	Greater than $100,000 in lost opportunity	Greater than 24 hours of consecutive unplanned downtime	At request of senior management discretion
Galv line	1 or more LTI	Notice of violation	Greater than $75,000 in equipment breakdown	More than 10 coils of the same consecutive defect	Greater than $75,000 in lost opportunity	Greater than 24 hours of consecutive unplanned downtime	At request of senior management discretion
Annealing line	1 or more LTI	Notice of violation	Greater than $75,000 in equipment breakdown	More than 25 coils of the same consecutive defect	Greater than $50,000 in lost opportunity	Greater than 24 hours of consecutive unplanned downtime	At request of senior management discretion
Temper mill	1 or more LTI	Notice of violation	Greater than $50,000 in equipment breakdown	More than 25 coils of the same consecutive defect	Greater than $50,000 in lost opportunity	Greater than 24 hours of consecutive unplanned downtime	At request of senior management discretion

(continued on next page)

Table 4.1 Example of an RCFA Minimum Trigger Point Decision Matrix (*continued*)

Asset Type	Safety	Environment	Repair Cost	Quality	Production Loss	Downtime	Other
Plating lines	1 or more LTI	Notice of violation	Greater than $505,000 in equipment breakdown	More than 25 coils of the same consecutive defect	Greater than $50,000 in lost opportunity	Greater than 24 hours of consecutive unplanned downtime	At request of senior management discretion
Inspection line	1 or more LTI	Notice of violation	Greater than $50,000 in equipment breakdown	More than 5 coils of the same consecutive defect	Greater than $75,000 in lost opportunity	Greater than 24 hours of consecutive unplanned downtime	At request of senior management discretion

We can use the following questions to guide us in our selection process:

- Was the failure caused by known asset design, capacity, or capability limitations?
- Was it caused by a known human error or action, procedure, or practice?
- Is there another specific, known reason for the failure to occur?

If the answer is "yes" to any of these questions, then we need to ask ourselves whether we know how to prevent this failure in the future. If we do, then we do not need to conduct an RCFA to tell us what we already know. If the answer is that we are *highly likely* to know, then we might want to conduct a simplified RCFA (described later in this chapter) and apply the rigorous approach only if this does not yield the desired results.

When starting on the journey to introduce RCFA in your department, site, or organization, you need to do so in a controlled manner. We recommend going slowly at the start in order to ensure that each analysis delivers the desired results. You can use these successes to incorporate RCFAs within your organization and then gradually increase the number of RCFAs performed, which ensures that each is implemented within a reasonable time frame and that you can track the associated benefits.

Collecting and Organizing the Information

It is often said that information is power. This is true for both troubleshooting *and* RCFA exercises.

When troubleshooting, the troubleshooter needs clear and precise facts right away (or as Leslie Neilson used to say, "Just the facts ma'am.") so that he or she can eliminate those that are not relevant and define possible solutions. There is usually an excess of information, much of which is not relevant to the problem, and the facts are too often hard to pin down. A good troubleshooter uses various methods, such as Socratic questioning (as discussed in Chapter 3), to gather the relevant information and organize it so that a strategy can be developed.

Information is also at the core of a good RCFA. While this is one of the key strengths of RCFA, too often people jump right to the analysis phase of the process, circumventing the collection and organization of the information. By "organizing" we mean the grouping and categorization of information in preparation for the eventual analysis.

There are a number of possible sources of failure information, and each requires its own particular tools for collection and organization. Data come from databases, spreadsheets, control charts, and reports. Meanwhile, observational information comes from interviews, process maps, and reports. Finally, experiential information comes from brainstorming exercises. In all cases, information collection is the act of gathering pertinent information about the failure and organizing it to facilitate the analysis of the failure, identify its root cause(s), and form a corrective strategy. While methods vary by source, the emphasis should be on ensuring that we accurately and comprehensively collect the information and organize it in a way that facilitates the analysis.

The most important point in collecting and organizing information is to know what we are using the information for. Once we have determined the purpose of the information, we can classify it in one of three groups:

1. Information for helping us to understand the situation:
    - This is what we use to check the extent of variability in our materials, processes, and products.
    - In simple terms, we want to understand: if the incoming material has changed (e.g.,

it's more granular, it's old and degraded, it's colder than normal, etc.); if we made changes to the process itself (e.g., changes in setpoint, temperatures, flows requirements, etc.); and if we made changes to the end product (e.g., quality requirements, packaging, etc.).

2. Information for analysis:
   - This is what we use in order to obtain correct information from the analysis (e.g., answering why this happened, when it happened, etc.).
   - We collect this information by examining our past results and testing new hypotheses.

3. Information for process control:
   - This is used to determine if the manufacturing process is normal (e.g., drifting or degradation of certain parameters, such as flow, output, current draw, etc.).

We then make adjustments and take action based on this information. Regardless of the information that we are collecting, we need to ensure that both the information that we gather and our means of collecting, organizing, analyzing, and comparing are in order to ensure that our information also answers our questions accurately.

Interviews and Reports

We want to gather failure information as soon as possible after the failure incident so that it is fresh in people's minds and as current as possible. This includes taking photos and sketches of the incident, collecting broken parts for future analysis (i.e., bag and tag), and interviewing the people involved. From this information we can generate a flash report (as we will see in Chapter 5) or an incident report. Both are generally used more for safety and environmental incidents, but they can and should be used for any failure that warrants an RCFA.

As part of the pre-analysis phase (i.e., collecting and organizing the information) we should also generate a preliminary investigation report. This may take many forms and should include a brief description of the problem, the response plan checklist (included in Chapter 5), evidence gathered to date, as well as any action items. This is a good memory aid when collecting the information and when writing the final report.

Another information gathering tool is a *process map*. This is when we use a workflow diagram in order to gain a clearer understanding of a process or a series of parallel processes. The mapping is generally done with those involved in the incident and/or who are experts in the process.

All information gathered from interviews, reports, and process mappings should be organized by type and relevance in order to facilitate the analysis stage. Interviews are used when there are no reports or the reports do not provide sufficient details. Process maps are used when a structure helps us to document our observations.

Data Collection and Classification

According to the *Merriam-Webster Dictionary*, "data" is "factual information (such as measurements or statistics) used as a basis for reasoning, discussion, or calculation."

In RCFA, we usually apply data collection and classification methods if the set of information is large and the data points are related. Data can come from a variety of sources:

- **Check sheets.** These help us to know the condition of the asset over a period of time. These could include laboratory results, operator logs, or security routes.
- **Control charts.** These are *statistical process control (SPC)* tools that we use to determine if a manufacturing process is in a state of control. SPCs are usually used to look at trends, such

as how often the pressure exceeded a certain threshold or the general tendency in the pressure level over the past 24 hours.

■ **Surveys.** For example, a list of safety-related incidents or the number of *work orders (WOs)* that exceeded the estimated time.

Data Classification Methods

Data is classified in preparation for the eventual analysis and is summarized in a table or graphic. The end result must give us insight regarding the incident, so that it is useful to us in the analysis phase. These can take the form of the following:

■ Frequency Distribution
■ Histogram
■ Pareto Diagram

Frequency Distribution   Frequency distribution is the simplest data collection method because it groups items by their frequency and provides a tally of each. A tally is the sum of how often a particular value is achieved. Frequency distribution is a tabular summary of data showing the number (frequency) of items in each of several non-overlapping classes. Frequency distribution is often the first step we take before we select a graphical approach to present the information. We use this method for both quantitative and qualitative data.

The following is a simple table that tracks the temperature of the product at the output of the centrifuge (Table 4.2a). High temperatures help the separation process but also accelerate the product coating the walls of the centrifuge and certain parts degrade more rapidly. Meanwhile, lower temperatures can lead to blockage. At this stage we are not analyzing the data, but rather we are simply recording and classifying it. The operators recorded the results.

Table 4.2a Frequency Distribution Example— Temperature Readings

	Temperature Reading (°F)						
Hour	Mon	Tues	Wed	Thur	Fri	Sat	Sun
1	105	116	100	122	113	123	113
2	115	123	123	112	123	116	123
3	108	105	116	123	115	109	113
4	98	111	121	111	112	119	111
5	111	117	114	127	116	117	127
6	119	105	127	115	113	115	112
7	112	124	113	112	104	127	119
8	121	116	121	117	127	106	116
9	116	103	128	115	116	121	102

Table **4.2a**   Frequency Distribution Example—
Temperature Readings (*continued*)

	Temperature Reading (°F)						
Hour	Mon	Tues	Wed	Thur	Fri	Sat	Sun
10	106	127	113	121	128	113	122
11	124	111	124	100	96	127	112
12	115	133	115	122	128	116	113
13	114	111	107	117	115	112	127
14	121	133	116	122	111	116	111
15	112	113	133	112	126	115	126
16	117	134	116	133	114	98	114
17	116	104	134	116	134	126	127
18	114	143	99	114	115	119	115
19	137	113	116	141	114	115	134
20	102	126	134	126	134	134	124
21	113	137	105	114	116	106	111
22	137	111	117	137	117	134	126
23	117	113	112	115	114	115	103
24	106	114	117	101	117	97	113

In order to construct a frequency distribution table (Table 4.2b), we need to divide the results (in this example, we have 24 samples per day for 7 days = 168) into intervals, and then count the number of results in each interval. In this case, the intervals would be temperature readings that fall within a 5°F range.

Table 4.2b   Frequency Distribution Table

Bin (Temperature range)	Tally	Frequency
<100°F	5	3.0%
100–105°F	9	5.4%
105–110°F	11	6.5%
110–115°F	43	25.6%
115–120°F	44	26.2%
120–125°F	20	11.9%
125–130°F	18	10.7%
130–135°F	12	7.1%
135–140°F	4	2.4%
>140F	2	1.2%

This table provides very interesting insights into the operating conditions of the centrifuge. You will note that the centrifuge temperature was 130°F or greater 10.7% of the time. This undoubtedly has an impact on the life expectancy of its parts and how often we need to rebuild it. You will also note that the centrifuge was running at 110°F or less 14.9% of the time. What impact does this have on the centrifuge's performance and final product quality? Does the colder product put stress on the equipment?

At this point, we do not know the impact of the temperature fluctuations but, thanks to the relatively simple data collection exercise, we now have data to analyze and possibly help us find the root cause and a solution.

Histogram A histogram is a graphical representation that helps us visualize the data compared to a table (as seen in the frequency distribution example in the previous section). Histograms are used in charting the precision of machines, in capability studies, when we examine causes that lead to changes in the process, and when we are analyzing the relationship between cause and effect.

When we create a histogram, we are creating a bar chart that shows how many data points are within a range (i.e., an interval). This is called a "bin." We usually choose the range that best fits our data. There are no set rules about how many bins we can have, but the rule of thumb is 5 to 20 bins. Any more than 20 bins and our graph will be hard to read, while fewer than 5 bins and our graph will have little (if any) value. Unlike a bar chart, the area of a bar in a histogram represents the frequency, rather than the height.

The following is a histogram of the centrifuge temperature example discussed previously (Figure 4.1).

This is the same data as the frequency distribution example. A histogram is usually easier to interpret and communicate, as the data is communicated graphically.

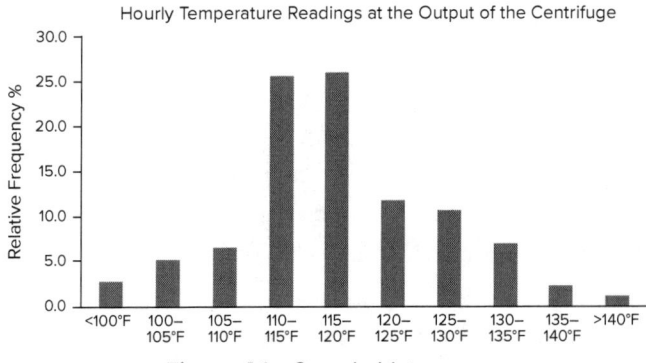

Figure 4.1 Sample histogram.

Pareto Diagram   Pareto goes a step further by graphing the information from the highest group of data (usually defects) to the lowest, providing a cumulative result as well. This helps us apply the 80/20 rule (also known as the "Pareto Principle") where we can focus on the 20% of defect groups that cause 80% of the failures. A Pareto diagram (Figure 4.2) is a type of chart that contains both bars and a line graph, wherein individual values are represented in descending order by bars, and the cumulative total is represented by the line.

The Pareto diagram shown in Figure 4.2 depicts the centrifuge temperature example discussed previously.

A Pareto diagram is a creative way of looking at the causes of problems because it stimulates thinking and helps us organize thoughts. These kinds of charts highlight

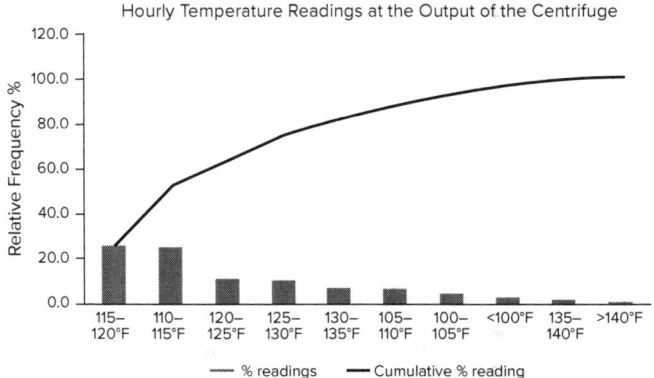

Figure 4.2 Sample Pareto diagram.

the most important among a (typically large) set of factors. A Pareto diagram indicates which problem we should solve first to eliminate defects and improve the operation.

In the case of the centrifuge output temperatures, we noticed that it was operating at between 110°F and 125°F 60% of the time, and outside of this range 40% of the time. Speaking with the process engineer, we discovered that the centrifuge should ideally be operated between 115°F–120°F; therefore, 110°F–125°F represents the outer range of acceptable temperatures. The Pareto diagram provided information that helped us to prepare for the analysis stage of the RCFA, where we will ask questions, like: "Why is the temperature too high?" and "Why is it too low?"

Choosing Between Methods

Oftentimes we use frequency distribution to collect and group the data, then create either a histogram or a Pareto diagram to organize it in preparation for the analysis. If we are looking to resolve the high frequency events, and we have failure data, we probably would use the Pareto diagram approach as it will help us apply the 80/20 rule. If we want to present the information in order (this would be useful in the centrifuge temperature example described in this section), then we would use a histogram.

Brainstorming Exercises

According to the *Merriam-Webster Dictionary*, "brainstorming" can be defined either as "a group problem-solving technique that involves the spontaneous contribution of ideas from all members of the group" or "the mulling over of ideas by one or more individuals in an attempt to devise or find a solution to a problem." In an RCFA, brainstorming almost always takes the former approach.

There are many brainstorming techniques available, from open unstructured approaches to very structured ones. The selection of a method is often based on our personal experience and training, as well as which method works best for us. For collecting and organizing information in order to conduct an RCFA, we recommend a structured approach such as the IS/IS NOT and Ishikawa techniques described in this section.

IS/IS NOT is designed to answer: "who," "what," "why," "where," "when," and "how" questions, while an Ishikawa diagram is designed to focus our thoughts on specific topics.

Brainstorming is a collaborative technique that we use to collect and organize information on a specific issue. This process needs to be spontaneous and a team effort in order to be effective. During the process, we employ an open discussion wherein the facilitator writes down all answers from the group. It is only when the initial brainstorming exercise is over that we review, validate, and rank the answers produced during this process. The brainstorming exercise may consist of one or many sessions, especially when additional information is required.

We use brainstorming exercises when the available data is not enough to help us understand the problem.

IS/IS NOT Method

A good brainstorming technique is the IS/IS NOT method (Table 4.3). The structure produces answers quickly and is an excellent memory jog that can generate constructive discussions. It is important to note that while using the IS/IS NOT method, we only answer the questions that are relevant to the incident under review. Further, the facilitator should encourage free, "out of the box thinking" and group discussions when answering the questions.

Table 4.3   IS/IS NOT Method for Brainstorming

	IS	IS NOT
WHO?	Who is affected by the problem?	Who is not affected by the problem?
	Who first observed the problem?	Who did not find the problem?
	To whom was the problem reported?	
WHAT?	What type of problem is it?	
	What device has the problem (part number)?	What does not have the problem?
	What is happening with the process and with containment?	What could be happening but is not?
	Do we have physical evidence of the problem?	What could be the problem but is not?
WHY?	Why is this a problem (degraded performance)?	Why is it not a problem?
	Is the process stable?	

Table 4.3 IS/IS NOT Method for Brainstorming (*continued*)

	IS	IS NOT
WHERE?	Where was the problem observed?	Where could the problem be observed but is not?
	Where does the problem occur?	Where else could the problem be located?
WHEN?	When was the problem first noticed?	When could the problem have been noticed but was not?
	When has it been noticed since?	
HOW MUCH?	Quantity of the problem (ppm—parts per million)?	How large could the problem have been but is not?
	How much is the problem costing in dollars, people, and time?	How big could the problem be but is not?
HOW OFTEN?	What is the trend (continuous, random, cyclical)?	What could the trend be but is not?
	Has the problem occurred previously?	

Ishikawa Diagram

An *Ishikawa diagram,* also known as a "cause and effect diagram" or a "fishbone diagram," is a graphical tool that helps identify, sort, and display possible causes of a specific problem. Though often thought of as an analytical tool, Ishikawa diagrams should be used to collect and organize information in a brainstorming session. The analysis comes later. The structure provided by the diagram helps team members think in a systematic way.

The following describes a typical five-step process used to develop an Ishikawa diagram (Figures 4.3a–c and 4.4a–d):

- **Step 1.** Identify and clearly define the outcome (the effect) that we want to analyze and create the "effect" box (Figure 4.3a). We state the effects as particular quality characteristics and problems resulting from the failure. In the RCM/FMEA world, this would be called a "functional failure."

Effect

Figure 4.3a Step 1 of the Ishikawa diagram.

- **Step 2.** Draw the spine of the chart and link it to the "Effect" box (Figure 4.3b). The spine is a horizontal arrow that points to the right. To the right of this line, the Effect box contains a brief description of the effect or outcome that results from the process.

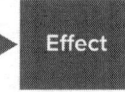
Effect

Figure 4.3b Step 2 of the Ishikawa diagram.

- **Step 3.** Identify the main causes of the effect that we are analyzing (Figure 4.3c). These causes are the labels for the major branches in our diagram, and they become categories.

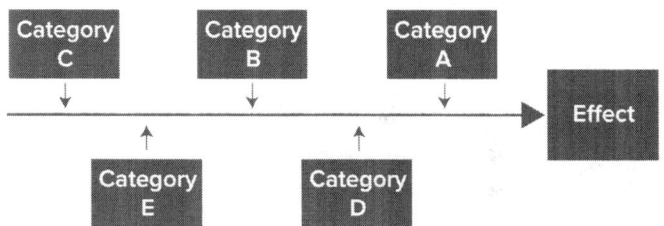

Figure 4.3c   Step 3 of the Ishikawa diagram.

It is at this point that we select the type of Ishikawa diagram that we will use. The primary ones include:

- 4S Diagram
- 8P Diagram
- Man, Machines, Materials Diagram
- Causes and Affinities Diagram

The *4S Diagram* (Figure 4.4a), is commonly used in the service industry because it considers surroundings, suppliers, skills, and systems—categories normally associated with this industry. That said, nothing prevents us from replacing or adding a category. For example, if the service is transportation of goods—even though equipment breakdowns should fall under systems—we might find it easier to create a fifth affinity called "Shipping" and keep the Systems affinity for computers and logistical systems.

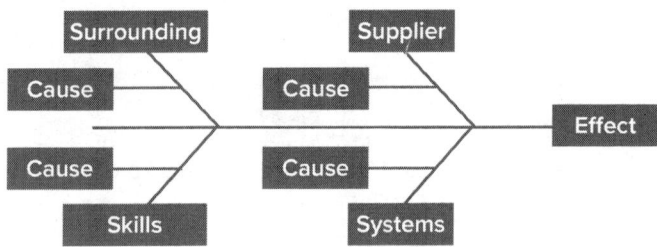

Figure 4.4a   Sample 4S diagram.

The *8P Diagram* (Figure 4.4b) can be applied in any type of business. In this case, we consider price, people, place/plant, procedures, promotion, processes, product, and policies. However, we might find this diagram a bit too broad when we are applying it to a specific equipment-related failure.

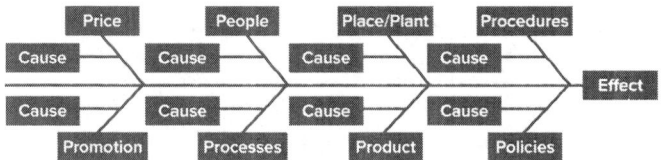

Figure 4.4b Sample 8P diagram.

The *Man, Machines, Materials Diagram* (Figure 4.4c) is commonly used in manufacturing and maintenance, and it is the most commonly used Ishikawa diagram overall. In this particular case, we consider man, materials, machine, methods, measurements, and environment. In some cases, we also include management/money and maintenance.

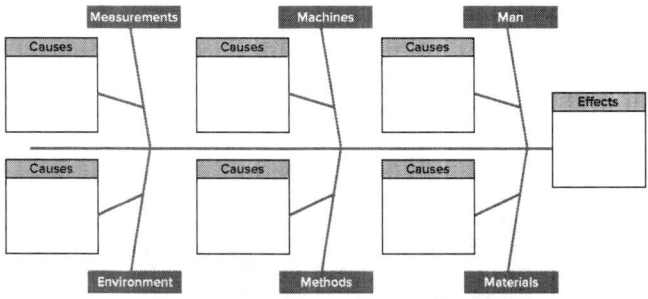

Figure 4.4c Sample Man, Machines, Materials diagram.

The *Causes and Affinities Diagram* differs from the earlier diagrams as the categories (affinities) are not predefined. In the Causes and Affinities Diagram, affinities are the category of causes (failure modes) (e.g., Process, Product, Policies). In its basic form, the Ishikawa Diagram has no predetermined affinities. We would select this model if the categories we initially selected do not match one of the predetermined models previously described. With the Causes and Affinities Diagram model, we select affinities that might be unique to our organization (Figure 4.4d). For example, a public relations firm might have affinities that would not be found in a manufacturing operation (e.g., political influences, social media). However, we recommend considering one of the predetermined models listed above and only using a Causes and Affinities

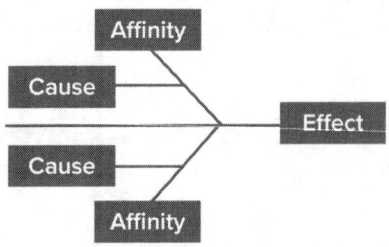

Figure 4.4d   Sample Causes and Affinities diagram.

Diagram as a last resort, as defining the affinities is an extra step that could take away valuable time from the brainstorming exercise.

■ **Step 4:** Once we have chosen a specific Ishikawa diagram, we regroup and start adding possible causes of the failure under each category. At this point there are no bad answers, and we do not need to fill each category evenly. If a cause applies to more than one category, then it can be listed under each (remember this is the brainstorming portion of the RCFA; we can deal with duplicates in the analysis portion).

■ **Step 5:** Identify increasingly more detailed levels of causes and continue organizing them under related causes or categories. We can do this by asking a series of "Why?" questions.

Once the Ishikawa diagram is filled out, we can proceed to the analysis stage of the RCFA.

Analyzing the Failure

In the previous section we focused on collecting and organizing the information necessary to conduct the analysis stage of the RCFA. In this section, we will discuss three possible approaches to conducting the analysis from the simplest approach, the *5-Whys* method, to *Logic Trees*, to finally a rigorous approach called *Single Functional Failure RCM (SFF-RCM)*.

It is important to note that the methods that we use for collecting and organizing data, such as the Ishikawa diagram, provide us with possible causes of failure but do not necessarily identify the root cause or the solution. Therefore, we need to apply one of the following methods to the output of these exercises in order to determine the right root cause and associated solution.

5-Whys

As the name states, the underlying approach of the 5-Whys is to ask: "Why did the failure occur?" until we determine the root cause, although we do not necessarily ask "Why?" five times. The goal of an RCFA is to determine the cause of a failure at the level at which we can anticipate it in order to take appropriate action to prevent it or mitigate its consequences in the future. Therefore, we ask: "Why?" as often as needed to determine this cause and its related solution. First, we start by clearly defining the problem. It is best to write this down.

For example, the observed effect (or problem) is that the conveyor stopped. Following the 5-Whys approach, we get a series of increasingly specific answers:

- **WHY?** Motor tripped on overload.
- **WHY?** Conveyor outlet is blocked.
- **WHY?** Product is too humid.
- **WHY?** There is a hole in the upstream asset's cover allowing water ingress.

In this case, after asking "Why?" four times, we determined that our solution is to find the hole(s) and patch it or replace the cover. Depending on the cause of the hole (i.e., wear, damage, or a modification), we might ask more "Why" questions and introduce additional strategies. For example, if it is from wear, and this has happened before, then we might make an addition to the periodic maintenance (PM) program to inspect for wear and possibly modify the equipment to minimize wear. If it is from damage, then we need to determine if that damage was preventable. We can look at physical barriers and guards for this answer. If the hole was generated during a modification, then we should look at our modification business process.

A 5-Whys exercise is a collaborative process and best conducted in a group setting with a trained facilitator. The group consists of subject matter experts from maintenance and production, and sometimes engineering, procurement, and the supplier. When facilitating a 5-Why exercise, the facilitator might find it difficult to start the process and keep it on track or may even need to stop it. For example, our plant is of a certain age where replacing all the assets with new ones would undoubtedly solve the water ingress issue—a typical assertion from participants when conducting an analysis. But this is not a reasonable solution due to the various associated costs to the business. Therefore, a facilitator must keep the group focused and aligned with the objectives on hand and keep the discussion practical.

The facilitator encourages introverts and manages extroverts to ensure that everyone who has something meaningful to say can say it. We recommend reviewing the Structured Questioning section of Chapter 3 for techniques to use while facilitating a 5-Whys analysis. In Chapter 5 we will discuss the skills of a good facilitator.

Logic Tree

Whereas the 5-Whys is sequential in nature, the Logic Tree can pursue multiple branches and possible causes of failure simultaneously (Figure 4.5). (The results of an

Ishikawa diagram will most likely feed into a Logic Tree diagram rather than a 5-Why analysis.) We can explore each branch to its sub-branches and their "root causes" by asking "Why?" and, eventually, cross off branches, sub-branches, and root causes as we deduce that they are not the cause of this particular failure. The Logic Tree approach provides us the opportunity to identify and manage multiple causes and potential causes of the failure during the one analysis. This is a more holistic strategy than the 5-Whys approach.

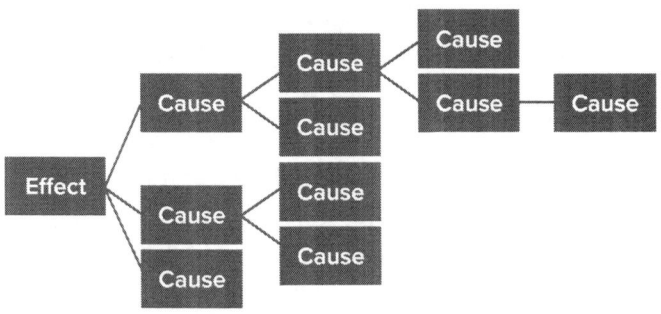

Figure 4.5   Logic Tree structure.

If we are not certain about the need to use a Logic Tree or simply a 5-Whys method, we recommend starting with the 5-Whys. If the resulting answers are not sequential (i.e., there are parallel answers), then we should continue the exercise with the Logic Tree method.

Logic Tree Steps

Conducting a Logic Tree analysis is usually quite different from a 5-Whys analysis because it can expand very rapidly. We can build the Logic Tree vertically or horizontally, depending on the group's preference and what provides the most visual information. We can use a whiteboard and draw boxes and/or use sticky notes. With the latter approach, we usually give each group member a sticky note pad so that they can write what they think. This approach makes it easier to move boxes around as the analysis progresses.

The following are key steps to follow when conducting the analysis:

- **Step 1. Describe the failure event (i.e., functional failure).** In this initial step, we need to ask: "In what way(s) does the equipment no longer do what its users want it to do?"
- **Step 2. Describe the failure effects.** This is where we determine what evidence exists that the equipment has failed to do what its users want it to do. For a protective device that has hidden consequences, we refer to its ability to act in the event of a failure. Describing the failure effects helps us to further define the failure.

■ **Step 3. Brainstorm the failure causes.** At
the start of this process we want to approach
the problem as a brainstorming exercise where
there are no wrong answers (it is easy to cross
these answers out later). When we are brainstorming, resist quick solutions, consider similar situations in other places, and ask why the
failure and its effects could have happened. The
facilitator should promote "out-of-the-box"
thinking to uncover all the credible potential
causes that could have led to the failure.

In our Logic Tree, we need to define the
first row of major causes or branches and add
the next row below this. Once the first two rows
of failure causes are identified, the facilitator
switches from a free-flowing exercise to a more
structured exercise using open and closed questions, especially asking "Why?" The facilitator
goes back to the first rows of causes of failure
and probes to see if any were left out and if all
causes are relevant. He or she continues from
there. The facilitator frames the brainstorming
into buckets and deals with one branch (or
bucket) at a time. Next, the group looks for
inter-relationships between the root causes,
as well as points of integration and adjusts the

Logic Tree accordingly. This is where sticky notes are useful because they allow us to make changes easily.

Failure cause categories include the following:

- **Equipment capability.** Under/over designed (upstream or downstream), etc.
- **Operating procedure or practices.** Loading, SOP, training, etc.
- **Maintenance procedures, programs, or practices.** Work instructions, PMs, training, tools, etc.
- **Process/product.** Product variations, selected production rate, quality requirements, etc.

For large RCFAs, after each analysis session, specific causes or lines of thought are assigned to a group member and tested or reviewed in the field to corroborate the group's thinking or collect more information. This is brought to the next analysis session and incorporated into the Logic Tree.

- **Step 4. Determine the physical root cause.** When the team has gone through the last set of potential causes and verified the evidence, they look at the equipment and review if all the reasonably likely physical root causes have been considered. The physical root cause might not be the failed asset itself, but what caused it to fail, such as mechanical fatigue, wear, degradation, alignment, etc. The root cause might be upstream or downstream from where the problem was observed.

■ **Step 5. Determine the human root cause.** The group then considers if all the reasonably likely human root causes have been addressed. These include errors in judgment, quality of the work, and in selection of parts or approach. Human errors can also include communication errors, not following procedures, and taking shortcuts.

The investigation into the human root cause is a search for the truth rather than a search for the "guilty." The goal is to identify the causes of failure, independent of their source, in order to determine a permanent solution.

■ **Step 6. Determine the latent root cause.** Most human errors take their origin from a system or process error. Therefore, after identifying a human root cause, we must consider if this was actually caused by a latent root cause. For example, we might ask: Was the WO sufficiently detailed? Were the parts properly identified? Was there a procedure for how to execute the work? Was the person trained? Is field tagging in place and is it correct? Was the person executing the work properly informed?

For example, we sent a mechanic to change a leaking steam valve during weekly maintenance when the powerhouse is shut down. The WO stated: "the top valve." The boiler house happens to have two top valves in this area, and neither are labeled. During normal operations, the leak is obvious, but during a shutdown, there is no steam and therefore no leak. Because of this, the mechanic changed the valve that looked the oldest. It was the wrong valve. This was not a human root cause—it was a latent one.

- **Step 7. Select a strategy (or strategies).** We conduct an RCFA to identify a solution to permanently resolve the problem, or at least mitigate its consequences. Once the Logic Tree is complete, we need to eliminate all causes of failure that do not lead to a permanent solution. What is left, hopefully, are causes that lead to a solution or multiple solutions. During Step 7 we need to define this solution(s) and develop a plan to implement and deploy the strategy. This will need to be presented to management for approval (Chapter 5 goes into more detail on this process).

- **Step 8. Prevent recurrence.** During this step, we have to consider related documents and systems (i.e., PM programs, health and safety manual, etc.) to ensure that these are updated based on the strategy (or strategies) selected in Step 7.

Logic Tree Example   The following is an example of a Logic Tree used to resolve the shortened life of the hammers and pins for a rendering plant's hammermills. The hammers are held in place by rods (pins). The mill rotates to grind product through contact with the hammers. Liquid fat is added in the upstream fluidizing tank to assist the grinding process. Previously, the hammers lasted four to six weeks, but in recent times this dropped to two weeks and even under one week in some instances.

We brought together a review team composed of maintenance, production, purchasing, and engineering. This team conducted an analysis over a number of sessions with research done between to address certain concerns or concepts. This included having the hammers and pins tested for their Rockwell hardness.

After discussions, we came to the conclusion that the causes of the issue fall under four main branches:

Process Causes, Product Causes, Manufacturing Causes, and Procedural Causes (Figure 4.6a).

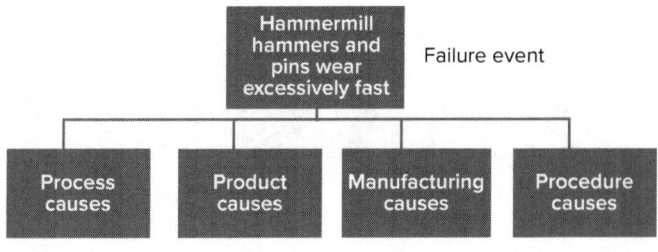

Figure 4.6a   Main branches of a diagnostic Logic Tree example.

In order to show the full logic tree in detail in this book, we broke it down to its four main branches below (Figures 4.6b–4.6e). When conducting an RCFA, we should never underestimate the amount of room we might need. For this particular example, we used a large white board and sticky notes. The board was dedicated for the activity and the analysis was conducted over a series of sessions. We filled the board.

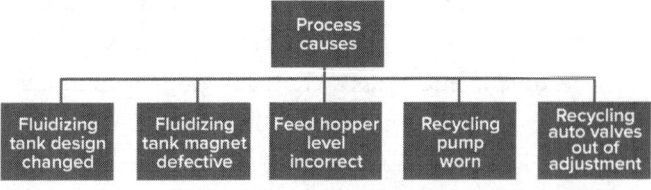

Figure 4.6b   Process causes branch of the Logic Tree example.

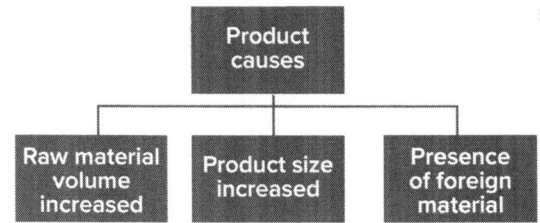

Figure 4.6c   Product causes branch of the Logic Tree example.

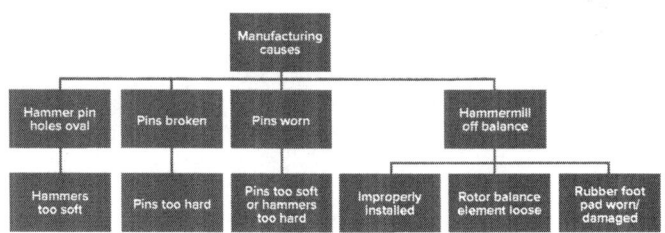

Figure 4.6d   Manufacturing causes branch
of the Logic Tree example.

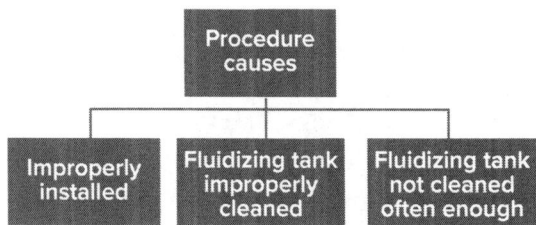

Figure 4.6e   Procedure causes branch
of the Logic Tree example.

The above charts depict the early stages of the analysis. The group added an extra level of causes in some cases and, after field validation, eliminated most of the initially listed causes. The final two main causes of failure were identified as:

1. Changes and inconsistencies in the hammer and pin harnesses
2. Changes to the fluidizing tank configuration

Both were addressed by the site in collaboration with the hammer and pin supplier and engineering. The presence of tramp metal in the fluidizing tank (this is then transferred to the "hammermill") is certainly another contributor to the issue and can be partially addressed through the cleaning process, though this requires line shutdown and would impact line availability.

Solution-Focused Logic Tree

There are two types of Logic Trees: Diagnostic Trees and Solution Trees. Diagnostic trees breakdown a "Why?" question and identify all of the reasonably likely root causes for the failure. Meanwhile, Solution trees break down a "How?" question and identify all of the reasonably likely issues that prevent the desired outcome. These trees also actively look for ways to resolve the issues.

It is important that whichever option we choose, we should not mix "How?" and "Why?" questions in the same Logic Tree. We always need to keep these questions separate as it will become difficult to keep the group focused on the activity. For example, let's say we want to reduce our car's fuel consumption and we start by asking why is the car consuming so much fuel? In the middle of the exercise, someone starts to describe physical changes one can make to the car to reduce fuel consumption (i.e., better tires or a tune up). We have now moved into answering the how question. In itself that is okay, but we may not have fleshed out all the why categories and causes and may therefore miss an important one. For instance, driving styles, weather, and type of fuel used can cause higher fuel consumption.

Consider this example: A copper refinery wanted to increase the time between shutdowns of their casting wheel from three weeks to four weeks. The reasons for the three-week interval were: greasing requirements, predictive maintenance (PdM) requirements, and cleaning requirements (in this case, copper splashing around the wheel during casting would eventually impede movement and damage sensors). The following is a simplified version of our Logic Tree for that copper refinery (Figure 4.7).

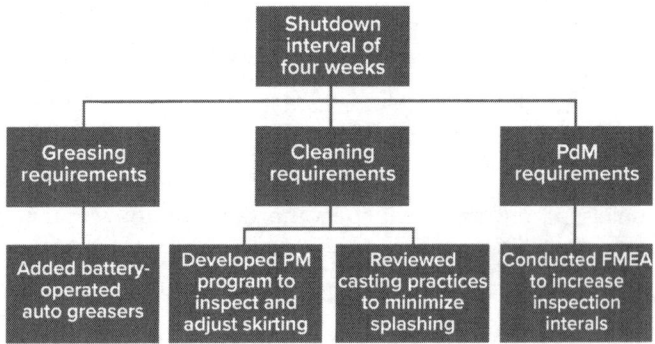

Figure 4.7 Solution-focused Logic Tree example.

On average, at the three-week interval, the wheel experienced six to eight unscheduled maintenance-related outages per month, and six to eight production-related outages per month. Due to operational requirements, if the wheel is stopped for more than five minutes, it has to be cooled down, cleaned, and reheated. This takes eight hours. Further, a shutdown is 12 hours. Therefore, the wheel was down for an average of 119 hours per month because of scheduled and unscheduled outages. This equals 16.3% downtime.

After we conducted the Logic Tree exercise, conducted the subsequent FMEA, and applied all of the recommendations, the site successfully moved from three-week shutdown intervals to four-week shutdown intervals. Additional benefits were achieved because we reviewed both operating practices and the maintenance program. As a result, the plant reduced the unscheduled outages to one per month due to maintenance and one to two per month due to production and introduced the four-week outage. This resulted in an average scheduled and unscheduled downtime of 33 hours per month, or 4.5%, which is a 12% gain in uptime.

Single Functional Failure RCM

Single Functional Failure RCM (SFF-RCM) is the third RCFA analysis methodology included in this book, and the most rigorous one. SFF-RCM takes its roots from RCM, the most rigorous analysis methodology available to develop preventive maintenance programs. RCM seeks to answer the seven questions, starting with "What do its users want the asset to do?" followed by "In what ways can it fail?" The latter is the asset's functional failures. The 5-Whys and Logic Tree methods call the functional failure the "effects" while, in the RCM and FMEA world, effects are distinct from the functional failure.

The Seven Questions of RCM
(from the SAE JA1011 Standard)

1. What are the functions and associated desired standards of performance of the asset in its present operating context (functions)?
2. In what ways can it fail to fulfill its functions (functional failures)?
3. What causes each functional failure (failure modes)?
4. What happens when each failure occurs (failure effects)?
5. In what way does each failure matter (failure consequences)?
6. What should be done to predict or prevent each failure (proactive tasks and task intervals)?
7. What should be done if a suitable proactive task cannot be found (default actions)?

As the name states, SFF-RCM focuses on a single functional failure, rather than all possible functional failures for the asset being analyzed. An SFF-RCM analysis can later be expanded to complete the application of RCM on the full asset.

An SFF-RCM is not an RCM, but rather, it is an RCFA. As such, it focuses on a specific failure that has already occurred and applies the in-depth information collection and organization elements described earlier in this chapter prior to beginning the analysis stage.

Why use SSF-RCM? RCM is conducted in a group and uses a decision diagram, or algorithm, when selecting a strategy to resolve the issue. The diagram distinguishes between safety, environmental, operational, and non-operational consequences and guides the group in identifying a solution that is both technically feasible and worth doing. Where this is particularly important is when the problem is very complex, and no single individual has sufficient knowledge to choose the optimal strategy. The RCM process provides a structured framework (more so than the previously discussed methods), so that, together, the group can identify a solution that would have been difficult to identify otherwise. The biggest difference lies in the use of a decision algorithm to guide the analysis of the failure in terms of consequences and recommended strategy. This removes the "emotion" from our decision process and encourages the selection of the most cost-effective strategy, one that is technically feasible and worth doing.

Following are examples of an SFF-RCM decision algorithm and worksheet, which are used to evaluate consequences of failure and select the appropriate action plan (Figure 4.8a and 4.8b).

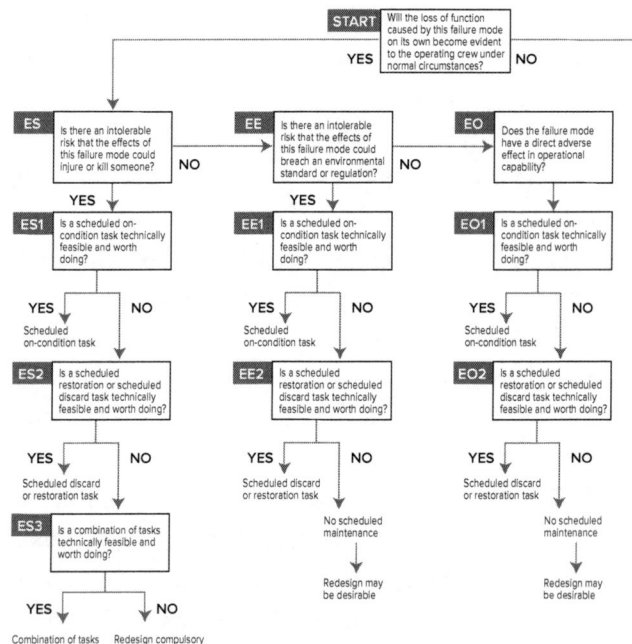

Figure 4.8a　SFF-RCM decision algorithm.

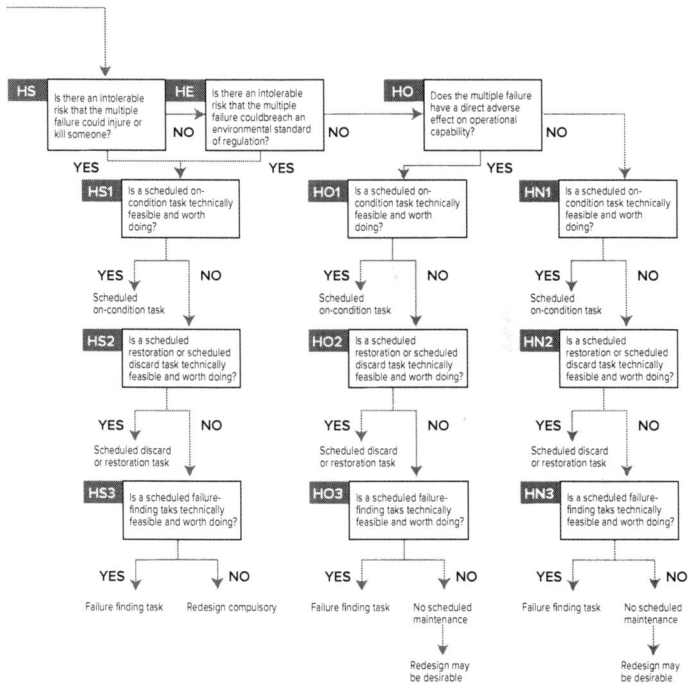

Figure 4.8a   SFF-RCM decision algorithm (*continued*).

119

Single Functional Failure RCM

| Asset | | Facilitator | | Date | |

| Function | | Functional Failure | |

					START	CONSEQUENCE				ACTION PLAN							
Failure Mode	Failure Effects	HIDDEN	SAFETY	ENVIRONMENT	OPERATIONAL	PREDICTIVE	PREVENTIVE	COMBINATION	FAILURE FINDING	Proposed Task	Initial Interval	Can Be Done By
1												
2												
3												
4												
5												
6												
7												
8												
9												
10												

Figure 4.8b   SFF-RCM decision worksheet (based on SAE JA 1012 example).

SSF-RCM is heavily structured and provides unparalleled guidance into the analysis and selection process. Therefore, SSF-RCM is the methodology of choice if the problem is complex, and there is insufficient knowledge to conduct a less involved methodology.

An added bonus of SFF-RCM is that beyond identifying the root cause, it can be used to address all other reasonably likely failure modes that could cause the selected functional failure.

SFF-RCM analyses are team exercises and require representation from maintenance, production, and sometimes engineering, as well as any other group that may have knowledge of this failure.

When conducting an SFF-RCM, analysis starts by defining the function affected. The function is what the asset's users want it to do. Don't forget to include at least one performance standard (gallons per minute, speed, temperature, volume, etc.). Next, define the specific functional failure that represents the problem that is being investigated by the RCFA review team. As the name states, this is the failure of the function that was initially defined.

From there, identify all the reasonably likely failure modes that could cause this functional failure. When defining the failure mode, include its effects. The following questions should be asked:

- Under what conditions does this failure occur (at start up, at shutdown, during normal operations)?
- What evidence is there that this failure is occurring or has occurred (alarm, noise, lower flow, line stopped, temperature drop, etc.)?
- Does this failure impact safety, the environment, operations?
- What happened/will happen if nothing is done to prevent the failure?
- What measures are needed to bring the equipment back to an operational state as well as how much time this will take, including production downtime?

Next, the group determines which failure mode or failure modes contributed to the functional failure, in this particular situation. This is the root cause of the failure. Using the RCM decision algorithm (Figure 4.8a), determine the consequence and an action plan to address this root cause(s). Use a worksheet to document the process (Figure 4.8b).

Once satisfied that the root cause has been identified and proper remedial action defined to address it, you may wish to continue the exercise for all the reasonably likely failure modes identified during this exercise. This will help resolve both the cause of this particular functional failure and possible future causes of this failure.

SFF-RCM Example

On an Ingersoll Rand XLE compressor in a steel plant, it was found that the low-pressure cylinder piston compression rings had completely worn away. Further, the aluminum low-pressure piston had seized inside the cylinder, the connecting rod between the low-pressure cylinder piston and the crankshaft was badly bent, and the crankshaft bearing between the low-pressure cylinder connecting rod and crankshaft was badly damaged.

Damaged and broken parts were collected, reviewed, and compared to new parts to better understand the failure. From this evidence, the review group determined that the most likely functional failure that would result in the low-pressure piston compression rings wearing so quickly was the loss of lubricating oil to the cylinder. Therefore, the functional failure to analyze was:

Unable to supply lube oil from the cylinder lube reservoir to the two injection points on the XLE air compressor low-pressure cylinder.

The next step in the analysis was to determine what caused the loss of lubricating oil to the low-pressure cylinder. The review group identified eight potential failure modes. As the failure effects for each failure mode were being described, it led to further inspections of the components of the cylinder lubrication feed system. The group found that the specific failure mode that caused the piston to seize in the low-pressure cylinder was:

> The solder securing the pump hold-down bar on the oil reservoir had fatigued.

The group decided that the oil reservoir hold-down mechanism should be redesigned so that the pump hold-down bar would be held in place with something other than solder. This modification was applied to all other air compressors with similar cylinder lubrication systems. The review group also recommended that a new protective circuit be designed and installed that would shut down the compressor in the event that any low- or high-pressure cylinder injection ports do not receive lube oil for more than one minute.

The next step in the process was to analyze each of the other seven failure modes and their associated effects and define the appropriate action plans to mitigate their consequences. Once implemented and deployed, the recommended actions resolved the problem and increased the compressor fleet's reliability.

Choosing Between Methods

When choosing between analysis methods, we recommend a few steps, beginning by determining if the cause and solution are already known and evaluating the level of risk to the organization. If we believe we know what to do and the risks are low, then we use a simplified approach. The 5-Whys is the most straightforward and easy to use analysis methodology, and also the least rigorous approach. Because of this, it lends itself well to a simplified approach.

If the risks are high, then we need to evaluate if there is sufficient individual knowledge within the proposed analysis team to successfully conduct an RCFA analysis. If the answer is "yes," then we use the Logic Tree approach. If the answer is "no," then we use the SFF-RCM approach.

This chapter summarized the RCFA methodology and explained how it differs from RCM, FMEA, FMECA, and PMO. We also introduced the different methods used to collect and organize the information in preparation for the eventual analysis and reviewed three RCFA analysis methods. The next chapter focuses on applying the RCFA techniques described in this chapter using a structured framework called the "Eight Disciplines." We will also discuss the role of the facilitator in conducting an RCFA and the skills of a good facilitator.

CHAPTER 5

Conducting an RCFA

So far, we have discussed understanding failures, trouble-shooting techniques, and the basics of RCFA. Founded in these ideas, this chapter will detail the RCFA in depth from pre-analysis to evaluating the results of the analysis. You can use this chapter as a guide for conducting an RCFA within your own organizations.

For the RCFA to be successful, it must follow a structured process and be facilitated. We will discuss facilitation and the skills of a good facilitator, as well as the business processes involved in RCFA.

After this, we will consider the steps of an RCFA in detail. This includes a look at the Eight Disciplines (8D) process, which stems from processes originated by the Ford Motor Company that have remained a standard in industry. This section will focus on our interpretation of the 8D process and include for each discipline (as well as the pre-analysis activities) sample documents that we created to help facilitate the RCFA and write the eventual final report. These examples will guide you from the pre-analysis activities through to the Eight Disciplines with alternate techniques that can be used for the information collection or management and analytical elements of the RCFA. We have also added examples for the deployment and for incorporating the business process required to successfully manage an RCFA analysis from initialization to completion and sustainability.

This chapter will cover:

- Facilitation
- Business Processes
- A Framework for Conducting an RCFA

Facilitation

A good facilitator helps the group achieve its goal by providing them unobtrusive guidance. In other words, they are focused on the RCFA *process*, while the participants in the team are focused on the *substance* of their work together.

It is important then that we implement a facilitator competency development program. This can be best compared to an apprenticeship for a tradesperson, in that this program needs to be hands-on and immersive rather than simply taking formal training sessions or courses. A facilitator competency development program needs to combine classroom theory and practical application in order for facilitators to become capable of applying the RCFA process efficiently and effectively.

The Role of the Facilitator

The basic assumption underlying group meetings is that two (or more) heads are better than one, and better decisions can be made if there is more input. However, to ensure that better decisions are made, the meeting must be facilitated.

The RCFA facilitator guides the organization and the RCFA participants in preserving, gathering, and organizing failure information; diagnosing this information; leading the root cause failure analysis; and developing an action plan that includes long-term prevention and sustainability of the results. During the analysis portion, the facilitator asks questions of a group of people chosen for their knowledge of the asset and the incident, ensuring that they reach a consensus about the answers, and recording these answers. As John Moubray said in his book, *Reliability-Centered Maintenance II*: "Of all the factors which affect the ultimate quality of the analysis, the skill of the facilitator is the most important. This applies both to the technical quality of the analysis and to the pace at which the analysis is completed and the attitude of the participants towards the RCM process." The same is true for RCFA.

The Qualities of a Good Facilitator

A good facilitator needs to have a specific combination of skills, personality traits, and adaptability in order to lead their team in RCFA. A facilitator competency development program needs to take into account these qualities both in terms of what the facilitator-in-training already has, and what he/she needs to improve. One of the most important is that this person is safety-minded. This includes broad knowledge of safety principles, as well as adherence to the organization's safety standards. The other key aspect is that this person understands the RCFA methodology fully, and can apply the concepts.

Part of being a good facilitator involves having good business skills. For example, during presentations the facilitator must sell him or herself (in other words they are competent and know what they are doing) as well as the RCFA process to the team. Therefore, the facilitator must understand and be able to quantify and articulate the impact that the incident has on the business. A facilitator also needs some level of business experience, such as knowledge of finances and ROI, in order to help the team select the right candidate event on which to conduct an RCFA and in order to measure success. Other skills include communication: a good facilitator listens and interprets information well; speaks and writes effectively; and can read and comprehend drawings, graphs, and charts. These communication skills are important because when the team knows the details of what is happening during the RCFA process, they feel that they are an important part of the team and become more invested in identifying a solution.

Because of the collaborative nature of RCFA, a facilitator needs to be able to think systematically in order to keep the team on track. This includes evaluating situations, solving problems, and making decisions. This requires that the facilitator focus both on the details and on the big picture. This person needs to understand the organization's goals and objectives at the system level, particularly because an RCFA is connected to all other activities within the organization. In order to balance these goals, the facilitator needs to be able to plan and manage time and resources effectively.

Another major part of RCFA facilitation is computer literacy. A comprehensive RCFA will include a significant amount of information and data—much of which will be digital. The RCFA facilitation, final recommendations, and report all require that the facilitator can use a computer, more specifically, the MS Office suite (Excel, Word, and PowerPoint), software tools such as the CMMS/EAM, and, possibly, the company's data historian.

Because the facilitator's role is as part of a team, this person needs to be good at working within a team culture. This involves coaching team members but also being patient with them. It also involves respecting people's differences and people's ideas and working to help the team come to a consensus when they are making a decision, yet doing so at a good pace. Part of working within a team is also being willing to learn and be coached. A good facilitator is open to change and can address when he/she is lacking in expertise.

In terms of the facilitator's interpersonal skills, this person needs to find ways to motivate the rest of the team. This involves the facilitator showing confidence in his or her decisions and in the team's ability to succeed. While this confidence is important, the facilitator needs to have the honesty and integrity to back it up. That is, this individual must have good credibility with peers and management and display professionalism. Specifically, the facilitator needs to uphold his/her promises and earn the team's respect rather than demand it. Additionally, the facilitator is responsible for motivating the team to get the job done, within the agreed upon timeline. This includes taking initiative, persisting in the face of difficult tasks, having a positive attitude even when the RCFA is not going according to plan, and encouraging the team to take needed breaks.

Business Processes

Too often, RCFAs are commissioned and executed but never implemented. This is because they were selected quickly in reaction to a problem and are not supported by a business process. That is, there is no process (or the process is not applied) to properly select RCFA as the analysis method of choice, in this instance, and for management to support the analysis and implementation of the eventual results.

As an entry point to an RCFA, we like to implement a business process that we call "Managing Failures and Actions" (Figure 5.1). This process includes the following sections:

- The recording or reporting of an incident (and what constitutes a reportable incident)
- Triage, which is co-managed by production and maintenance. That is, both need to be involved in choosing RCFA as the right strategy to resolve the problem, and both need to support the subsequent activities

- Assigning the appropriate group with responsibility for the action. The chosen group develops a plan
- Plan approval by the appropriate area of the business
- Control and management, which ensures that the recommended actions are implemented in a reasonable time
- Sustainability tracking

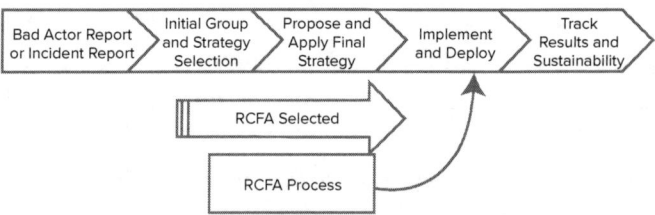

Figure 5.1 Managing failures and actions.

Reportable incidents or bad actors are failures of an asset, component, or process that has a significant adverse impact on one or more of the following: safety, the environment, or economic factors. Bad actors can result from both chronic and sporadic failures, although they are more likely to be chronic failures. These should be reported formally, either using an incident report or a *bad actor report (BAR)*.

If RCFA is selected, we need to consider how detailed we will go with our analysis (Figure 5.2). If the solution is well known and/or the risks are low, a simplified RCFA might be appropriate. Otherwise, we need to employ a detailed approach. In both cases, the approved recommendations must be implemented and deployed, the results tracked, and elements put in place to ensure sustainability of the results.

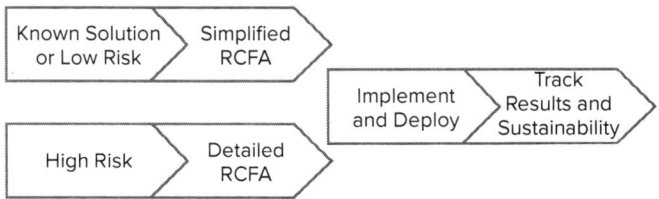

Figure 5.2   Selecting between a simplified and a detailed approach.

The following map depicts the RCFA business process (Figure 5.3a). It incorporates a simplified and a rigorous approach, and identifies the role of the different resources in conducting the RCFA and in implementing the results. The subsequent diagrams focus on specific areas of the process map (Figures 5.3b–5.3d).

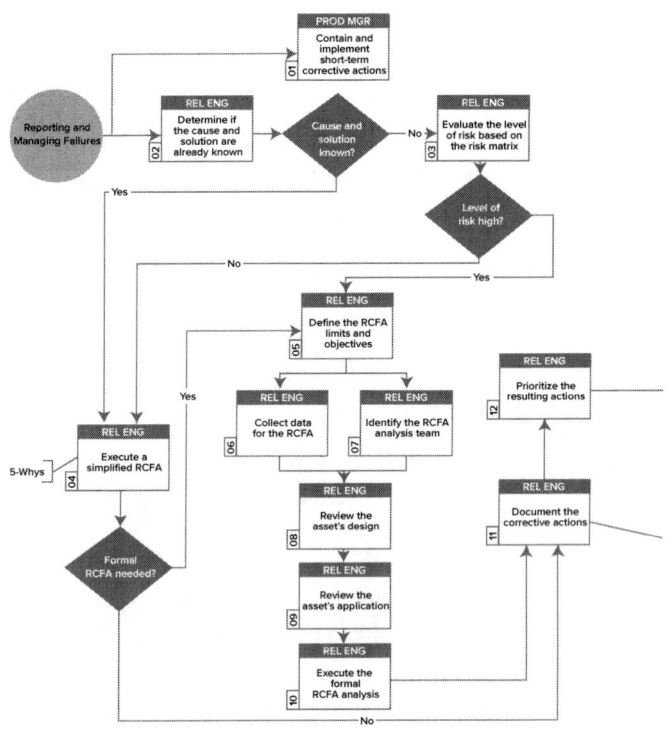

Figure 5.3a   The RCFA business process.

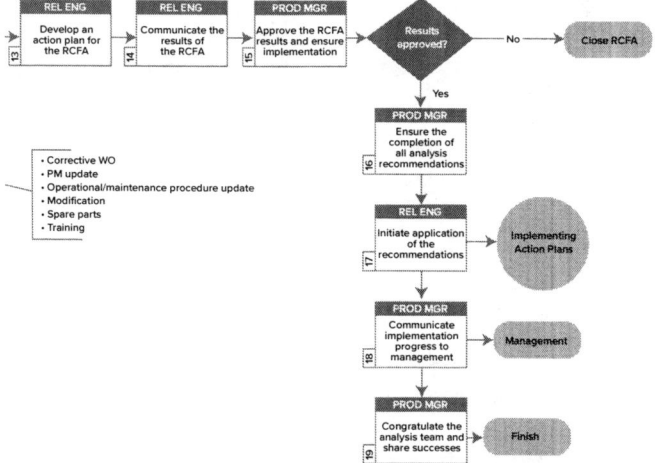

Figure 5.3a   The RCFA business process (*continued*).

A feed into the RCFA process is the Reporting and Managing Failures process (discussed in Figure 5.1). From there we determine if the cause and solution are already known and if the risks are high. This leads us to choosing between a simplified or a rigorous approach. In parallel, we ensure that the problem is contained and implement short-term corrective actions.

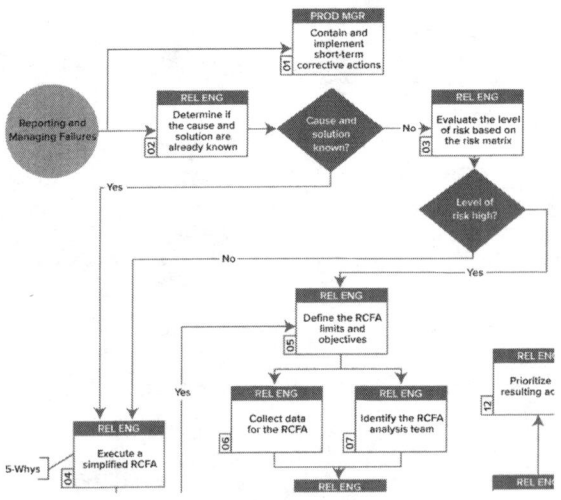

Figure 5.3b RCFA business process— reporting and managing failures.

If a rigorous approach is selected, then information is collected, a formal group is convened, the failure incident is analyzed, corrective actions are identified, an action plan is developed, and the results are communicated.

Even though 5-Whys is the method of choice for the simplified approach, it can still be used for the rigorous approach. The main difference is the level of preparation done prior to the analysis phase. In all cases, an action plan must be drafted and communicated (Figure 5.3c).

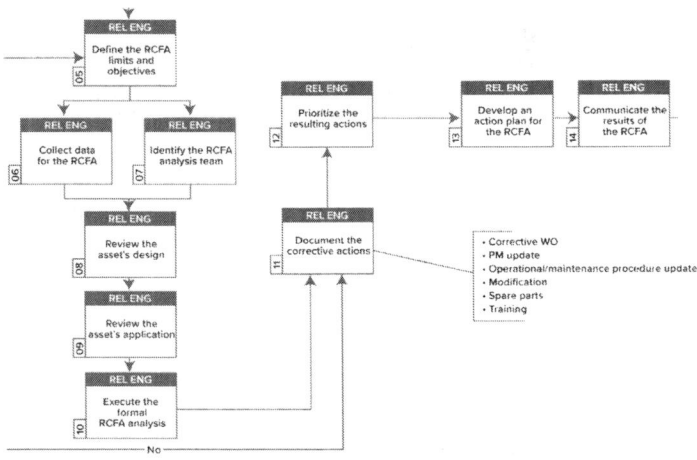

Figure 5.3c   RCFA business process—
developing an action plan.

At this point, you will notice an interesting change for most roles and responsibilities: in a traditional RCFA environment, the reliability specialist is responsible and accountable for the whole process, including reporting progress. The production manager is the asset owner and, in our process, he/she approves the results and is accountable for the implementation and deployment of the results, as well as for reporting the progress to management (Figure 5.3d).

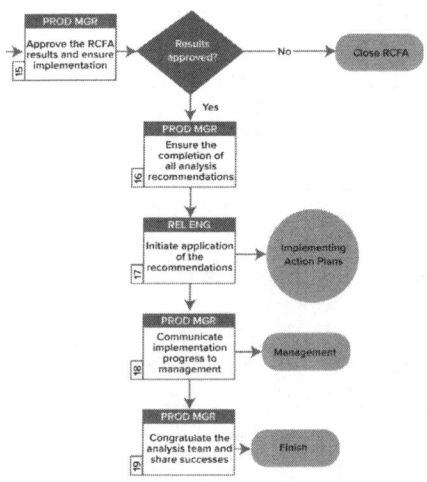

Figure 5.3d RCFA business process—approving the results.

A Framework for Conducting RCFA

In the previous section, we explored the elements and practices that are used to conduct an RCFA exercise. Here we will look at a framework to structure the process so that it is efficient and effective. In this case, a major component is the Eight Disciplines (8D), developed by the Ford Motor Company in the 1980s. It was designed to provide structure when identifying the root cause of a problem, devise a short-term fix, and implement a long-term solution to prevent recurring problems. The 8D is not an RCFA methodology; rather, it is a framework to guide us in leading and managing the RCFA.

Pre-Analysis

The pre-analysis activities include developing a plan, responding to the incident, applying a response plan, and preserving and recording evidence—including any preliminary reports.

Develop a Plan

Before applying the 8D process, we need to plan. The process is rigorous and should only be applied when appropriate. The planning process is an analysis of the work to be performed during and after an RCFA and the merits of applying the RCFA. The plan includes a description of the problem and its impact on the organization, the resource requirements, and the timeline. This should be presented to the asset owner for approval before starting the analysis.

The 8D framework is useful because it results in not only a problem-solving process, but also a standard and a reporting format. Before we begin, we need to ask ourselves: "Does this problem warrant or require an RCFA?" If so, then we need to explain why before we proceed.

To create structure and traceability, we need to write down the following: a short description of the issue, a tracking number, who authorized this analysis, and whether an emergency response action is needed or was undertaken.

Respond to the Incident

The first pre-analysis step that we need to take prior to the RCFA is dealing with the results of the failure. Our top priority at this point should be safety and the environment, including caring for injured personnel, bringing the facility to a safe condition, and stopping or cleaning any environmental spills or loss of containment. If we are using RCFA in an incident investigation, our response to an emergency should still focus on bringing the facility to a completely safe condition and to clean or contain any environmental issues.

We need to prioritize, in order, personnel safety and rescue, firefighting, environmental containment, and preventing further equipment damage. If anyone is injured or could get injured, we need to take care of them prior to dealing with any evidence. We also need to have a documented RCFA response plan based on the severity of the incident, including a safety and loss control process.

In safety and environmentally conscious organizations, incidents AND near misses are reported and acted upon.

Apply a Response Plan

The next pre-analysis step is having a response plan. An RCFA response plan should include:

- Safety of personnel, the environment, and the general public
- Maintaining physical facilities
- Incident escalation process
- Response plan to return the failed equipment to service
- Collection of information relative to the incident
- Media response kit

The following is an example of a response plan checklist (Table 5.1). It is important to note that this is only a *sample* and that you need to build a checklist for your own personal organization and possibly for each particular context or department.

Table 5.1 Example of a Response Plan Checklist

Response Plan Checklist	Yes	No
Safety of Personnel, Environment, and the General Public		
Procedure exists		
Procedure posted in key areas		
Personnel trained		
Procedure applied adequately		
Maintaining Physical Facilities		
Shutdown procedures exist		
Procedures in place to ensure viability of the assets		
Procedures posted in key areas		
Personnel trained		
Procedures applied adequately		
Incident Escalation Procedure		
Procedure exists		
Company contact(s) listed with phone number(s)		
External contacts listed with phone numbers		
Procedure posted in key areas		
Personnel trained		
Procedures applied adequately		
Response Plan to Return Failed Equipment to Service		
Plan exists		
Plan is easily accessible		
Personnel are trained		
Plan is applied adequately		

Table 5.1 Example of a Response Plan Checklist (*continued*)

Response Plan Checklist	Yes	No
Collection of Information Relative to the Incident		
Procedure exists		
Procedure posted in key areas		
Personnel trained		
Procedure applied adequately		
Media Response Kit		
Kit exists		
Kit is easily accessible		
Personnel are trained		
Kit is applied accurately		

Preserve and Record Evidence

The sequence of events that happened before a failure might be crucial in determining the root cause. Because of this, we need to preserve as much evidence as we can before we conduct an RCFA and thoroughly study the evidence as soon as we can after the failure event occurs. It is also important that everyone in the organization is aware of their role in case of a major failure. If the event is less serious in nature, there might be time to assign a trained individual to the task.

Evidence collection is particularly crucial with sporadic failures because the team might only have one chance to collect the critical information that they need to perform an RCFA. Depending on the degree of the incident, the area may have to be cleared until the investigation is complete. This would be the case with a fatality or a major fire.

First, we need to deal with the results of the failure event. It is unlikely that everything can be left as is, but we need to remember to preserve and record the evidence as much as reasonably possible. For repeat failures, we should put plans in place to collect or trend failure information prior to the failure event. We need to develop information collection strategies based on the frequency of occurrence and the consequence of the failure event. For example, at one of my clients they were experiencing frequent failures of one specific pump, while the other two similar pumps and applications were surviving much longer. At first glance, the installations, application, and pump types were the same (we even swapped pumps with no measurable change). To resolve this, we had to track conditions across the three pumps over an extended period of time. This helped us identify the root cause of the problem (the inlet pipe configuration was different) and make appropriate changes.

When we are preserving or recording evidence, it can be helpful to use an evidence preservation checklist such as the "5 Ps" (Table 5.2), as shown next.

Table 5.2   Example of an Evidence Preservation Checklist

Type	Checklist Items
Parts	List parts.
	Have photos been taken?
	Have parts been bagged and tagged?
	For chronic failures where the parts are hidden (such as internal components), we need to carefully collect evidence while repairs are underway. A member of the investigation team should witness the uncovering of evidence, when possible, or ask the repair facility to carefully document the state of the components, using photographs and diagrams.
	For sporadic failures, the area should be roped off so that parts and samples can be carefully collected in their found state after positions are noted. Process sampling (gases, liquids, solids) should be collected as necessary. We also need to "bag and tag" parts after recording positions. However, we should not try refitting parts back together or reinstalling them because it might contaminate the evidence.
Position	Have the position of people and parts been photographed?
	Have sketches been made?
	Another important piece of evidence is the position of all people and parts at the time of the incident. These positions should be photographed or videotaped, when possible, or sketched showing what was found where. We should also note the positions of valves, switches, indicators, people, and equipment. The position of people is more often required when there has been a safety incident.
People	Who should be interviewed?
	Ask the 5Ws: Who? What? Where? When? Why?

(continued on next page)

Table 5.2 Example of an Evidence Preservation Checklist (*continued*)

Type	Checklist Items
	Sometimes people fear that they will be implicated in an incident and therefore sabotage evidence or efforts to determine root cause. Because of this, we need to make it clear that the intent of the investigation is not to discipline employees but to prevent repeat failures.
	When possible, interviews should be conducted privately and individually as soon as possible after an event has occurred. Eyewitnesses that were involved prior to or during the failure event can supply the most crucial evidence. We need to determine questions prior to conducting interviews so that we can be prepared and as unbiased as possible.
	We also need to list the key people to be interviewed and why, including what type of information they will provide. The minimum interview list needs to include the following: the person who did the field troubleshooting, the operator on shift the day of the failure, whoever knows the design of the operational process and loading, anyone who did follow-up work, and whoever is the specialist on that particular equipment.
Paper	These include: Operator log Alarm printouts Process and equipment loading strip charts and readings Laboratory records Inspection records Equipment failure investigation form

(continued on next page)

Table 5.2 Example of an Evidence Preservation Checklist (*continued*)

Type	Checklist Items
	Field paper documentation can be crucial when we are conducting an RCFA. We need to collect relevant paperwork as soon after a failure as possible. This applies to such things as process condition trends. Timing is important because some information is purged from the system after a period of time.
Production	What was the production loading?
	Our main focus with production is checking to see the production loading (tons, percentage of target, etc.) at the time of the incident, including settings, readouts, and trends in production parameters. Again, some of these are purged regularly and should be collected as soon as possible.

Another useful tool when preserving or recording evidence is an equipment failure investigation form, such as the one in Figure 5.4. This particular document is extremely useful because it lists all the possible cause categories, which can help to jog the person's memory. As with previously mentioned documents, it is important for you to adapt this form to your specific operating context (Figure 5.4).

Equipment Failure Investigation

Date/Time of Failure: _____
Failure Description: _____
Business Area: _____
Department: _____

Instructions: Operating Supervisor or Technician needs to fill out this form and take pictures of the failed equipment as soon as possible after the failure occurred with the people involved during the failure and repairs (Production and Maintenance personnel).

A. Nature of Failure: (Mark with an X)

Arced	Bent	Binding	Blown	Blunt	Twisted
Brinelled	Broken	Burn	Burst	Compressed	Warped
Corroded	Cracked	Cut	Dirty	Delaminate	Tripped
Flashed	Grooved	Leaking	Missing	Melted	Worn
Over-heated	Perished	Pitted	Scored	Seized	Torn
Sheared	Stretched	Stripped	Stuck	Other	

(continued on next page)

Figure 5.4 Example of an equipment failure investigation form.

Did the failure start in one place and progress across an area or length? Yes ☐ No ☐

Were equipment failure pictures taken and sent to the Reliability Team? Yes ☐ No ☐

Where is the failed equipment stored for further investigation?

Full Failure Description: (include any contributory events leading up to or causing the failure)

_____
_____
_____
_____
_____
_____
_____
_____
_____
_____
_____
_____
_____
_____
_____
_____
_____
_____
_____
_____
_____
_____
_____
_____
_____
_____
_____
_____
_____
_____
_____
_____

(continued on next page)

Figure 5.4 Example of an equipment failure investigation form (*continued*).

B. Environmental Conditions During Failure—Indoor or Outside: (Mark with an X)

Normal		Dust		Heat		Cold		Steam	
Smoke		Vapors		Water		Other			

C. Prefailure Warning Signs: (Mark with an X)

Normal		Heat		Smell		Vibration		Performance	
Out of spec		High amps		NONE		Other			

Remarks: (provide details on items marked with an X)

_____

_____

_____

_____

_____

D. What Changes Occurred Before the Failure?

D. Changes in Equipment Loading: (Mark with an X)

Physical load		Stress/pressure		Temperature	
Electric current		Electric voltage		Flow rate	
Product handled		Other			

D. Changes in Ambient Conditions: (Mark with an X)

Gas composition/amount		Heating/cooling		Shock/vibration levels	
Liquid composition/amount		Other			

D. Changes in Process/Action/Function: (Mark with an X)

Change in process		Process setting changes		Different operator	
Maintenance intervention		Other			

Remarks: (provide details on items marked with an X)

_____

_____

(continued on next page)

Figure 5.4 Example of an equipment failure investigation form (*continued*).

Why and when did the above changes occur?

_____

_____

_____

E. Identification of Causes and Contributory Conditions: (Mark with an X)
Equipment failure occurred because something went wrong in one of the following processes:

Specification		Design		Manufacture		Strorage	
Handling		Installation		Operation		Inspection	
Maintenance		Other					

People: Human error may have contributed to the equipment failure. This may have happened at the machine or along the equipment or material supply chain.

Insufficient training		Improper procedure	
Lack of concentration		Intent (sabotage)	
Other			

Remarks: (provide details on items marked with an X)

_____

_____

_____

F. What did the failure cost?

Safety/injury		Environmental compliance		Operational delay	
Repair cost/time		Major part consumption			

Form completed by:

_____   _____   _____
Name (Printed)   Signature   Date

Figure 5.4 Example of an equipment failure investigation form (*continued*).

Preliminary reports are also excellent sources of information because you can use them to collect information from those present before details are forgotten. A common tool is an Incident Flash Report, as shown in Table 5.3.

Table 5.3 Example of an Incident Flash Report Form

Incident Flash Report Form	
Basic Information of the Event (Use Safety Incident Report if Safety-Related)	
Name	
Title	
Date and Time of Incident	
Incident Type	
Location of the Event	
Department	
Location	
Details	
How did the incident occur?	
Was there risk of personal injury?	
Environmental damage?	
Property damage?	
Secondary damage?	
Equipment operational state	
Was evidence preserved?	
Any photos taken?	
Describe the incident	
Date	Reporter's signature

The 8D Process

The 8D Process starts once the pre-analysis activities are completed and consists of eight disciplines (or "8Ds"):

- D1 Establish the Team
- D2 Define the Problem
- D3 Develop Interim Containment Action
- D4 Identify Root Cause
- D5 Identify Permanent Corrective Actions
- D6 Implement and Validate
- D7 Prevent Recurrence
- D8 Recognize Team Effort

D1: Establish the Team

The team consists of the facilitator and the participants. We should always identify a *champion* or *sponsor*. Though this person might choose to not participate in the actual analysis sessions, he/she will play a vital role in the eventual validation and implementation stages and must stay involved or informed of the progress and results.

The facilitator is the team leader that is trained in the RCFA methodology. He/she should not be expected to know every detail about a particular failure. However, his or her job is to facilitate the process by bringing in experts from the field, technical support, laboratories, and vendors to help put the pieces of the failure analysis into place. The facilitator keeps bias and preconceived ideas of the failure cause (that might result in wrong conclusions) from interfering with the brainstorming sessions.

The makeup of the cross-functional RCFA team will vary depending on the failure being analyzed. A team can range from about three to eight people, with additional outside resources brought in as needed. Team members often include an operator, mechanic, electrician, various specialists, and vendors. At least one person on the team should be unaware of the failure events in order to promote "out-of-the-box" thinking. When additional information is required, the facilitator assigns team members tasks, such as gathering parts, recording positions, conducting interviews, and collecting documents.

The following table is useful when selecting the team members (Table 5.4). We should always identify what skills they bring to the analysis as well as their responsibilities. This will help garner support for their participation from themselves and their superiors. At this point, it is also important to define the team objectives.

Table 5.4   Establishing the Team

| Department | Subject Matter Experts | | |
	Name	Skills	Responsibilities
Team Objectives			

D2: Define the Problem

Defining the problem is one of the most important elements of a good RCFA process. This might include brainstorming exercises (i.e., an IS/IS NOT analysis), data grouping (i.e., histograms and Pareto charts), and causal diagrams (i.e., Ishikawa diagrams). This step sets the

RCFA exercise apart from most RCM, FMEA, FMECA, and PMO analyses because this is part of the collecting and organizing the information in preparation for the eventual analysis of one specific failure and would be too demanding when building or optimizing a PM program for a complete asset or production unit.

The following are examples of graphs and tables that might be useful when defining the problem (Figures 5.5a–5.5d). These techniques were discussed in Chapter 4.

	IS	IS NOT
WHO?	Who is affected by the problem?	Who is not affected by the problem?
	Who first observed the problem?	Who did not find the problem?
	To who was the problem reported?	
WHAT?	What type of problem is it?	
	What device has the problem (part ID)?	What does not have the problem?
	What is happening with the process and with containment?	What could be happening but is not?
	Do we have physical evidence of the problem?	What could be the problem but is not?
WHY?	Why this is a problem (degraded performance)?	Why is it not a problem?
	Is the process stable?	
WHERE?	Where was the problem observed?	Where could the problem be observed but is not?
	Where did the problem occur?	Where else could the problem be located?
WHEN?	When the problem was first noticed?	When the problem could have been noticed but was not?
	When has it been noticed since?	
HOW MUCH?	Quantity of problem (ppm)?	How many could have the problem but don't?
	How much is the problem costing in dollars, people, and time?	How big could the problem be but is not?
HOW OFTEN?	What is the trend (continuous, random, cyclical)?	What could the trend be but is not?
	Has the problem occurred previously?	

Figure 5.5a   IS/IS NOT table example.

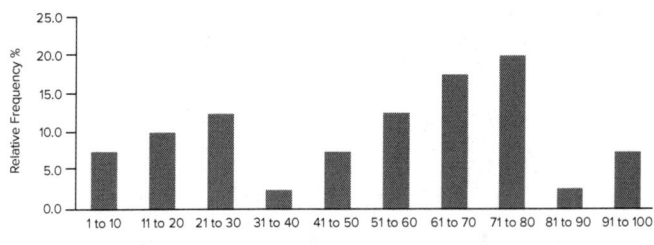

Figure 5.5b Histogram example.

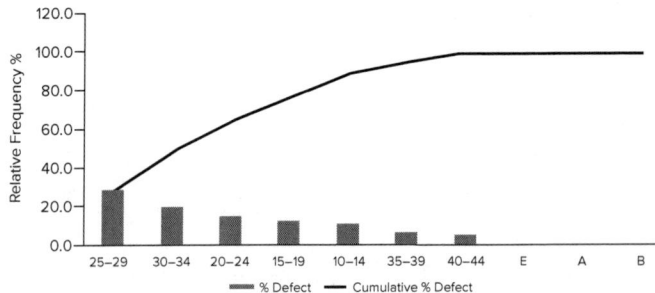

Figure 5.5c Pareto diagram example.

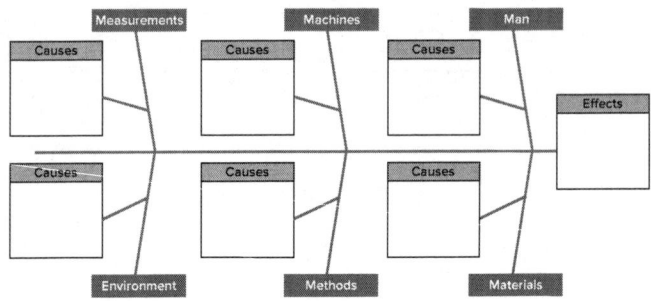

Figure 5.5d Ishikawa diagram example.

D3: Develop Interim Containment Action

This includes actions to contain and temporarily resolve the problem until a permanent solution is in place. It is important that we make a formal list of these actions in order to ensure that we have not missed anything important and that all identified actions have been validated and are completed. This is a weak area for most organizations because they rarely take the time to write it down. As a result, important ideas are missed or forgotten, there is no follow up on some actions, and communication is poor.

One of my clients went through a bad production period where lots of equipment kept failing (poorly maintained or improperly used) and they faced constant quality issues. Experts were brought in, and some of the problems went away, but overall, the plant was still limping along. Eventually, a friend of mine took over the situation, brought everyone together, wrote down the issues and action plans, the group make informed decisions and an interim plan was developed and implemented. He had the courage to stop, take time to plan, and then act. Roughly 36 hours later, the problem was under control and much of the solution identified and applied. As the plant was now in control, the rest of the organization had time to analyze the situation and implement permanent strategies.

We recommend setting up a brainstorming session with all key stakeholders and with subject matter experts

(Table 5.5). Consider safety and environmental requirements when defining this action plan. The plan must ensure that any safety or environmental risks are removed, and it must not create new ones. We need to assign an owner for the Interim Containment Action Plan. This person ensures that the actions are executed and has the group reconvene if additional actions are required.

Table 5.5 Example of a Brainstorming Tool for Interim Containment Actions

Action	Validated

D4: Identify Root Cause

RCFA is a logical approach for determining physical, human, and system (latent) root causes for any type of failure. During the analysis, we need to list and verify the operational route cause(s), the physical root cause(s), and the human error root cause(s), including those that are system-, process-, or procedure-based.

In Chapter 4, we presented three methods that can be used to identify the root cause(s) of the problem (Figures 5.6a–5.6c):

- **5-Whys.** This is the simplest of the methods. We ask "Why?" as often as needed to identify the cause of failure at the level at which we can anticipate to take appropriate action.

Describe the functional failure	
Ask—Why did this happen?	
Ask—Why did this happen?	
Ask—Why did this happen?	
Ask—Why did this happen?	
Ask—Why did this happen?	
Ask—Why did this happen?	
Ask—Why did this happen?	
Ask—Why did this happen?	

Figure 5.6a   A 5-Whys worksheet.

- **Logic Trees.** This is similar to the 5-Whys method in that we ask "Why?," but the biggest difference is that we do so on parallel fronts at the same time, creating branches and offshoots of the initial question. Logic Trees can be used to solve a failure by asking "Why?" or find solutions to a requirement by asking: "How?"

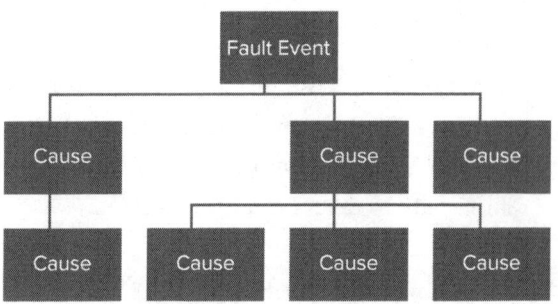

Figure 5.6b A Logic Tree.

- **SFF-RCM.** This is the most rigorous method and should be used for complex failures when there is not enough individual knowledge to solve the issue and the group needs the RCM structure to uncover the root cause. However, this should not be confused with RCM; this is an RCFA methodology.

Asset			Facilitator			Date	
Function							
		Functional Failure					

Failure Mode	Failure Effects	Consequence Evaluation											Default Tasks			Proposed Task	Initial Interval	Can Be Done By
		H	S	E	O	H1 S1 O1 N1	H2 S2 O2 N2	H3 S3 O3 N3					H4 H5 S4					
1																		
2																		
3																		
4																		
5																		
6																		
7																		
8																		
9																		
10																		

Figure 5.6c An SFF-RCM worksheet.

D5: Identify Permanent Corrective Actions

The RCFA is only valuable if it brings about a permanent solution to a problem. This might include changes to the PM program, physical modifications, modified or new procedures or practices, enhancements to the training program and job descriptions, and capital expenditures.

The following table lists typical types of solutions identified during an RCFA (Table 5.6). For each action selected, we must identify a due date and assign responsibility for its execution.

Table 5.6   Permanent Corrective Action (PCA) List Template

Proposed Task	Due Date	Responsible
Proposed Modification	Due Date	Responsible
Proposed Procedure	Due Date	Responsible
Proposed Safety/Environmental Practice	Due Date	Responsible
Proposed Operating/Maintenance/ Materials Management Practice	Due Date	Responsible
Proposed Training Program	Due Date	Responsible
Proposed Hiring Practice	Due Date	Responsible
Proposed CapEx	Due Date	Responsible

D6: Implement and Validate

It is crucial that we implement and deploy the results from the RCFA in a reasonable time frame so that the site can achieve the projected benefits, the effort put into the analysis is not wasted, participants see value in their efforts, and the site is encouraged to conduct future RCFAs. Part of this involves communicating our results and recommendations to the rest of the team. We need to review these findings with the appropriate personnel, which usually includes meetings with management. The team should have a consensus on the root causes of the incident, the recommendations, and the projected benefits prior to issuing the report.

It is important that we plan properly if we want to successfully implement a permanent change. Activities should include developing a project plan for implementation, including steps to implement and deploy, communication, success metrics, and sustainability, communicating that plan to all stakeholders, and measuring the improvements in order to validate them.

Even the best RCFA is useless if it is not implemented and deployed. The following checklist is an example of the steps to include in a project plan (Table 5.7). We recommend that you develop one for your site and even one that is specific to the RCFA being conducted.

Table 5.7 Example of a Project Plan Checklist

Activity	Due Date	Responsible	Status
Develop/Implement			
First field validation			
Manage modifications			
Physical modifications			
Procedures/work instructions			
Adjust hierarchy as needed			
Implement actions in CMMS			
Second field validation			
Manage repairs			
Conduct five question review			
Can the task be done safely?			
Do we have easy access?			
Are things properly labeled?			
Do we need special tools or materials?			

(continued on next page)

Table 5.7 Example of a Project Plan Checklist (*continued*)

Activity	Due Date	Responsible	Status
Would a visual aid help the task?			
Develop and add visual aids, tags, and tools as needed			
Approve program			
Develop deployment strategy			
Develop/adapt training programs			
Deploy			
Conduct awareness sessions for front line supervisors			
Deploy new program through front line supervisors			
Add program to morning meeting and KPI board			
Collect feedback and adjust as needed			
Sustain			
Audit results			
Coach front line supervisors in coaching the people involved			
Follow KPIs and adjust program as needed			

D7: Prevent Recurrence

Once we have found a solution to the problem and implemented it, we want to consider what other systems and assets would benefit from this solution. To this end, the group lists similar systems for review and, if appropriate, implementation of the solution. We also want to make certain that the solution is not causing other problems. Therefore, the group (or a subset of the group) reviews existing documents and systems and identifies changes (if any) that are required as a result of the applied solution. Finally, the group reviews and introduces KPIs and audits as needed.

The following are a check sheet template for addressing similar systems and assets (Table 5.8a) and check sheet example for reviewing documents and systems (Table 5.8b). Again, please adapt to your specific needs.

Table 5.8a Address Similar Systems
and Assets Checklist—Template

Process/Item	Responsible	Due Date

Table 5.8b Review Documents
and Systems Checklist—Template

Document/System	Responsible	Change Required?	Due Date
Health and Safety Manual			
Environmental Compliance Manual			
Quality Manual			
Management System Manual			
Work Instructions			
Practices and Procedures			
Process Flow Charts			
Process Control Plans			
PM Program			
Training Matrix			
Job Descriptions			
KPIs			
CMMS Hierarchy			
Risk and Criticality Assessment			
Materials Management Strategy			
Other			

D8: Recognize Team Effort

A final step in the RCFA exercise (prior to implementation) is to recognize the team and ask for their feedback. This will help optimize the process for the next time as we are focusing on our successes as much as our failures. Team members need recognition and a chance to provide feedback.

Table 5.9 is an example of a process feedback form that can be used.

Table 5.9   Process Feedback Form

Was this problem-solving exercise effective?		
Yes/ No	Signature/ Title/Date	Comments/Findings/Lessons Learned

The Final Report

After we have completed the analysis, we still need to write a report. This will be used to inform management of the steps taken and results achieved and will probably be necessary to receive approval to spend effort and money in implementing and deploying the results. The report also helps create a culture change towards conducting RCFAs and provides information for the future, as our production process changes, and as we face new challenges.

Beyond writing the report, we also need to think of where we will post it, as part of our communication plan, and store it for future reference. We probably have an internal system, such as SharePoint—this is a good place to store the report. Alternatively, we can store it in our CMMS/EAM. Because the majority of the RCFAs will be conducted to solve equipment-related failures, the CMMS/EAM (with its physical hierarchy) is a good place to store the report for future retrieval.

The following shows a template to write a *FRACAS (Failure Reporting and Corrective Action System)* report—one of the more popular reporting structures (Table 5.10).

Table 5.10 FRACAS Report Form

Revision	Revision Date	Reviewed	Approved
Date		Report Number	

General Information	
Type of Equipment	
Location of Failure	
Team Members	
Equipment Number	
Date of Failure	
Time of Failure	
Total Downtime	

Description

Problem History

Action Plan Summary (See attached action plan for details)

Root Cause Analysis Diagram

Table 5.10   FRACAS Report Form (*continued*)

Action Plan			
Task	% Complete	Responsible	Due Date

Comments

Photographs

Maintenance Manager   Production Manager

Example RCFA Report—Boiler Ash Conveyor

The boiler ash conveyor's chain stopped working; after investigation, we found out that the hydraulic motor was damaged internally. To better understand the issue and fina a permanent solution the company decided to conduct an RCFA. This is the resulting report.

General Information	
Type of Equipment:	Boiler Submerged Ash Conveyor
Location of Failure:	Hydraulic Motor
Equipment No:	282-075
Date of Failure:	3/23/2014
Time of Failure:	16:00
Total Downtime:	12 hrs

(continued on next page)

Description

The conveyor's chain stopped working; after investigation, we found out that the hydraulic motor was damaged internally (unable to use the hydraulic pressure and convert it to work).

History of Problem

We experienced several blockages during the last week on that conveyor, due to different causes, including a huge amount of ash falling on the chain and metallic parts interfering with the chain. When these blockages occurred, the hydraulic pressure went very high before it stopped. Several "jog" tests were then performed when attempting to restart the system. On our last blockage, we were unable to restart the chain; we then found out that the motor had failed.

Root Cause

Preliminary analysis shows that failure was due to motor internal leakage, probably due to recent high pressures and hits ("jogs").

Action Plan Summary

1. Replace the motor with a new one (done)

2. Clean the conveyor (done)

3. Send the motor to a repair shop for investigation and repair (results to come)

4. Adjust action plan if required according to motor failure analysis results

5. Investigate recent changes that may have caused last week's blockages

6. Investigate the material entering the power boiler & ash getting out of it

Logic Tree

Logic Tree used to conduct the analysis.

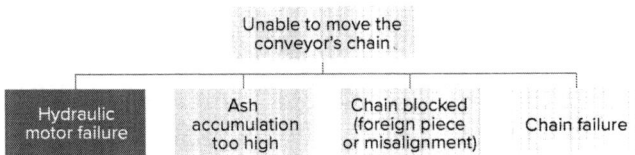

Detailed Comments
The motor repair shop people confirmed that the motor's cylinders where worn out and the clearances where too high. That explains why the motor wasn't able to deliver enough torque to operate at high loads.
We will adjust preventive maintenance on this motor and we will monitor the oil leakage to evaluate the cylinder wear.

Example RCFA Report—ID Fan

Issues with a turbine's ID fan were investigated and resolved using an RCFA.

General Information	
Type of Equipment:	ID Fan Drive Turbine
Location of Failure:	Trip mechanism
Equipment No:	461-211
Date of Failure:	8/9/2014
Time of Failure:	9:40 PM
Total Downtime:	660 minutes

Description
Trip mechanism suddenly released, stopping steam flow to the turbine. It was manually reactivated, tripped again minutes later. After a few tries, mechanical team investigated and found that the carbide parts were worn and the lever not correctly aligned.

History of Problem
During the Annual Outage, we replaced our turbine with a refurbished one. During commissioning, we found issues with its internal gear clearance. Since this problem couldn't be fixed in the allowed outage time, the old turbine was installed. After the Outage, the old turbine started to trip without reaching an over speed or low oil pressure. The first out recorded for this event was a mechanical trip. It was difficult to restart due to its trip mechanism tight tolerances. Same events happened during the evening of August 9. It was restarted, but tripped again moments later, and so on. After several tries to keep it running, the mechanical team went in for investigation and repairs. We had to stop another similar turbine (non-critical to operations) for parts. But even with the good parts, it was difficult to restart due to alignment.

Action Plan Summary
1. Order new parts to rebuild the old trip mechanism
2. Reinstall the refurbished trip mechanism on the other non-critical turbine
3. Replace the trip mechanism with a new one during the next outage
4. Modify startup procedure to ensure validation with a mirror that both sides of the carbide parts are fully engaged

Logic Tree

Detailed Comments
This trip mechanism has very tight tolerances and very little interference between its two main parts that are in contact.
We need to ensure an optimal contact between these two parts, or it will disengage over time, due to vibrations and changes in speed or load.

Evaluating the Results

Once the RCFA has been completed, we need to periodically evaluate its results. This entails observing and assessing the results against the objectives that we established at the beginning of the analysis. If the objectives are not met, or the results start to slide, then we need to revisit the analysis and make the appropriate course corrections. From both a plant reliability and change management perspective, it is important that we evaluate the results of the RCFA in order to help maintain organizational awareness and ownership in the process and the results.

Because many of our objectives will be financial, we need to track downtime, quality losses, and excessive costs. The following are a few key definitions that can be useful for doing so:

- **Overall Equipment Efficiency (OEE)** refers to availability × quality × machine speed. In simple terms, OEE is the measure of good product produced compared to how much we could have produced if everything were correct (i.e., if we can produce 1,000 good units per hour and only produce 800 due to stoppages, quality losses and the speed we are running the equipment, OEE is 80% for this line). Increases in asset management efficiency and effectiveness translate into increased asset availability, better quality results, and the possibility of increasing machine speed without penalizing the first two. This results in an increase in production throughput.

- **Conversion Loss Reductions (CLR)** include the excess use of energy, water, and consumables because of inefficiencies in the process. These losses are caused by excessive consumption due to poorly functioning equipment and/or consumption while the equipment was not used. CLR also includes the quality loss component of OEE (the benefits must not be counted twice). Quality losses include both the scrapping of the product and, when applicable, the cost to rework the product.

- **Maintenance Cost Reductions (MCR)** are the maintenance labor and spare part consumption benefits that can be achieved from an increase in maintenance efficiency and effectiveness.

- **Indirect Cost Reductions (ICR)** include operator overtime, excessive spare parts inventory, demurrage charges and penalties, work in progress, excessive finished goods inventory levels, and capital expenditures. Some of these are directly impacted, but most require additional steps to achieve the benefits.

Objectives of the RCFA can also include safety, regulatory compliance, and fewer environmental incidents. We need to track these and apply course corrections as needed.

Finally, an important deliverable from an RCFA exercise is cultural change. We can use surveys or stakeholder analyses to measure the impact of the RCFA exercise on the team and specific team members or groups. It is not unusual that—despite our best efforts—some people are not satisfied after the exercise is complete, implemented, and deployed. A simple and effective remedial action to resolve this is communication. We can revisit the results, post them, and even lead awareness sessions to keep people informed.

In order to be successful, we need to clearly define the approach that we will use to evaluate success and communicate this to the key stakeholders. The management team must be onboard with our chosen approach and actively sponsor it.

This chapter focused on applying the RCFA techniques described in Chapter 4 using a structured framework called the "8 Disciplines." We discussed the role of the facilitator in conducting an RCFA and the qualities of a good facilitator. This chapter covered how to achieve consistent results in applying the RCFA process and how to make RCFA part of your company culture. We also discussed how to evaluate the results, once the exercise is complete.

CHAPTER 6

Conclusion

Problem solving is both a science and an art. We need the tools, techniques, and structure to apply both immediate (troubleshooting) and longer-term/permanent (RCFA) methodologies to resolve failure incidents, or at least mitigate their consequences. We also need people with the aptitude for problem solving, and we need an organizational structure and culture that trains and supports these individuals.

This book introduced us to the many elements of good troubleshooting and RCFA and showed us that to consistently achieve our goals, we need to select the right techniques and see them to term. Problem solving takes time and effort, and there are no shortcuts if we want to achieve sustainable benefits.

In Part I of this book, we focused on what constitutes a failure as Chapter 2 discussed our understanding of failures, what they are, their causes, and types of failures. Part II, which is comprised of Chapter 3, discussed when to troubleshoot, asking structured questions, the process itself, the qualities of a good troubleshooter, and what to do after the troubleshooting activity. Together, these gave us a strong foundation for moving into Part III: Root Cause Failure Analysis (RCFA).

In the second half of this book, we used problem-solving and troubleshooting principles throughout our discussion of RCFA. Chapter 4 focused on the RCFA methodology and explains how it differs from RCM, FMEA, FMECA, and PMO. Meanwhile, Chapter 5 focused on applying the RCFA techniques described in the previous chapter using a structured framework called the "8 Disciplines."

In addition to understanding these concepts and practicing these methods, we also have to have the courage to fail. We conduct troubleshooting and RCFA activities because we do not know or understand the issue and want to identify the cause and solution. There is no guarantee that we will be successful the first time. As individuals and organizations, you need to trust and follow the process and take the time to do so properly. By following a process and applying the techniques and methods described in this book, you will be consistently successful in resolving your site's failure incidents.

This brings us to the business case for problem solving. There must be value added to invest the effort and cost to conducting these activities; otherwise, your organization will invest this time and money somewhere else. You are good at identifying and measuring downtime, but what about uptime, reduced safety and environmental incidents, quality gains, and customer satisfaction? Your troubleshooting and RCFA KPIs should include cumulative metrics to track the benefits these processes bring to your organization, as well as the effort and cost expended to apply them. You also need to measure and report on how many have been implemented and deployed and how many have achieved (or not achieved) their objectives.

Establishing problem-solving processes, tools, and competencies is an essential element of a successful asset management-focused organization. The return on investment is high, and the benefits continue long after the exercise is finished.

Glossary

List of Acronyms

8Ds	8 Disciplines
BAR	Bad Actor Report
CLR	Conversion Loss Reductions
CMMS	Computer Maintenance Management System
EAM	Enterprise Asset Management
FMEA	Failure Mode and Effects Analysis
FMECA	Failure Mode, Effects, and Criticality Analysis
FRACAS	Failure Reporting and Corrective Action System
ICR	Indirect Cost Reduction
KPI	Key Performance Indicator
MCR	Maintenance Cost Reduction
MTBF	Mean Time Between Failure
OEE	Overall Equipment Efficiency
PMO	Preventive Maintenance Optimization
PM/PdM	Preventive Maintenance/Predictive Maintenance

RCM	Reliability-Centered Maintenance
RCFA	Root Cause Failure Analysis
RCA	Root Cause Analysis
SFF-RCM	Single Functional Failure Reliability Centered Maintenance
SPC	Statistical Process Control
WO	Work Order
WR	Work Request

Key Definitions

80/20 Rule	A rule that we follow for focusing on the 20% of failures that cause 80% of the problems. Also known as the Pareto Principle.
8D Process	A structure for conducting an RCFA analysis originated by the Ford Motor Company in the 1980s. The 8 Disciplines are as follows:

D1 Establish the Team
D2 Define the Problem
D3 Develop Interim Containment Action
D4 Identify Root Cause
D5 Identify Permanent Corrective Actions

	D6 Implement and Validate
	D7 Prevent Recurrence
	D8 Recognize Team Effort
5-Whys	A method of asking "why" the problem occurred until we arrive at the root cause. (Note: This does not necessarily have to be five times.)
Cause Resolution	Focused on determining the cause of the failure so that a permanent solution can be identified and, if approved, implemented.
Chronic Failure	These failures are generally not dramatic or difficult to repair and they usually occur gradually. They tend to be repetitive or short-term, affecting production and/or maintenance. They are also accepted as an operational cost.
CMMS/ EAM	Computer software that is designed to simplify maintenance management.
CLR	These include the excess use of energy, water, and consumables due to inefficiencies in the process. These are caused by poorly functioning equipment and/or are consumed while the equipment was not used. CLR also includes the quality loss component of OEE (the benefits must not

be counted twice). Quality losses include both the scrapping of the product and, when applicable, the cost to rework the product.

Emergency or Break-in Work Work that cannot be planned or scheduled.

Failure Effect Describe what happens when a failure occurs and nothing is done to prevent it. The effects provide enough details so that we can choose the right strategy to mitigate the failure consequences.

Failure Mode A term used to describe any event that causes a failed state, listed at an appropriate level of detail. These descriptions consist of a noun and a verb (e.g., "bearing worn" or "cable damaged"). Failure modes can also be called "causes of failure."

FMEA Developed in the late 1950s to improve military equipment reliability, it involves reviewing assets at the component level to identify reasonably likely failure modes and the strategies necessary to eliminate or minimize their failure consequences.

FMECA Similar to FMEA, with the exception that it involves criticality when we are deciding if the task is worth doing.

Frequency Distribution	A tabular summary of data showing the number (frequency) of items in each of several non-overlapping classes
Histogram	A graphical representation that helps us visualize the data. They are used in charting the precision of machines, in capability studies, when we examine causes that lead to changes in the process, and when we are analyzing the relationship between cause and effect.
Human Cause	Anything caused by improper human action. This can include a deviation from intention, expectation, or desirability.
ICR	These include operator overtime, excessive spare parts inventory, demurrage charges and penalties, work in progress, excessive finished goods inventory levels, and capital expenditures. Some of these are directly impacted, but most require additional steps to achieve the benefits.
Ishikawa Diagram	A graphical tool that helps identify, sort, and display possible causes of a specific problem. Though often thought of as an analytical tool, these diagrams are used to collect and organize information in a brainstorming session. The analysis comes

later. Also known as a *cause and effect diagram* or a *fishbone diagram*.

IS/IS NOT Method   A structured questioning that drives quickly to answers and serves as a memory jog that can generate constructive discussions. It focuses on the 5Ws and "how."

KPI   Quantifiable measurement that reflects the critical success factors of the organization.

Latent Cause   System- or process-related errors.

Logic Tree   Rather than a sequential method (such as the 5-Whys), this approach provides us the opportunity to identify and manage multiple causes and potential causes of the failure during the one analysis.

MCR   These are the maintenance labor and spare part consumption benefits that can be achieved from an increase in maintenance efficiency and effectiveness.

OEE   Refers to availability × quality × machine speed. Increases in asset management efficiency and effectiveness translate into increased asset availability, better quality results, and the possibility of increasing machine speed without penalizing the first two. This results in an increase in production throughput.

Pareto Diagram A tool for graphing information from the highest number of defects to the lowest, providing a cumulative result as well.

Physical Cause These are usually some type of mechanical or electrical failure. It might not be the broken component itself but rather what caused it, such as mechanical fatigue or a specific type of corrosion from a particular source. It might be upstream or downstream from where the problem is manifested.

PMO This is similar to FMEA in that it defines the failure modes and identifies appropriate strategies to mitigate their failure consequences. However, PMO is almost always used to review and optimize existing programs rather than for developing new programs.

RCFA A method that we use to determine the cause of a failure once it has occurred at the level at which we can anticipate it in order that we can take appropriate action to prevent future failures or mitigate their consequences.

RCM Derived from Stan Nowlan and Howard Heap's research on the US airplane industry in the 1960s, it is a process used to determine what must be done to ensure that any physical asset continues to do what its users want it to do in its present operating context. The goal of RCM is to mitigate the consequences of a failure either by avoiding it or by reducing its consequences (if the consequences are significant).

SFF-RCM This method focuses on a single functional failure rather than all possible functional failures for the asset being analyzed. An SFF-RCM analysis can later be expanded to complete the application of RCM on the full asset.

Sporadic Failure Failures that usually occur infrequently, suddenly, and dramatically, and lead to significant consequences.

Trouble-shooting Focused on resolving the effects quickly so that we can go back to normal, or near-normal, operations.

Index

About the Author

J. R. Paul Lanthier, P. Eng., is a facilitator, practitioner, trainer, coach, mentor, project manager, practice lead, Director, VP of Operations, and President. Paul has worked in the fields of asset management, organizational engineering, reliability and maintenance at all levels in various industry sectors worldwide. He is the former Director of "The Aladon Network," and leader for RCM2, as well as former Director of Reliability Services at Ivara Corp.

As a professional engineer with nearly 40 years of experience, Paul is a recognized technical leader in the field of asset management and an author of multiple books.

Asset Management and Reliability are technical, tactical, and strategic exercises that consider human and organizational requirements in order to ensure sustainability. Paul has championed efforts for many organizations in this field to ensure that the approaches are holistic, sustainable and practical.